Modelling Interactions Between Vector-Borne Diseases and Environment Using GIS

Modelling Interactions Between Vector-Borne Diseases and Environment Using GIS

Hassan M. Khormi
Umm AL Qura University, Makkah, Saudi Arabia

Lalit Kumar
University of New England, Armidale, Australia

CRC Press
Taylor & Francis Group
Boca Raton London New York

CRC Press is an imprint of the
Taylor & Francis Group, an **informa** business

CRC Press
Taylor & Francis Group
6000 Broken Sound Parkway NW, Suite 300
Boca Raton, FL 33487-2742

First issued in paperback 2021

© 2015 by Taylor & Francis Group, LLC
CRC Press is an imprint of Taylor & Francis Group, an Informa business

ISBN 13: 978-1-138-59723-5 (pbk)
ISBN 13: 978-1-4822-2738-3 (hbk)

Library of Congress Cataloging-in-Publication Data

Khormi, Hassan M., 1984-
  Modelling interactions between vector-borne diseases and environment using GIS / Hassan M. Khormi, Lalit Kumar.
      pages cm
  Summary: "This timely and groundbreaking book demonstrates how to develop models of vector borne disease risks based on different environmental and socioeconomic variables and to assess the association between these variables and their vectors in a Geographic Information System (GIS) environment. It addresses new spatial approaches and techniques based on location and environment and introduces methods to identify, determine, and analyze the trend, movement, and distribution of diseases and the vectors that transmit disease"-- Provided by publisher.
    ISBN 978-1-4822-2738-3 (hardback)
    1. Communicable diseases--Mathematical models. 2. Communicable diseases--Epidemiology--Mathematical models. 3. Communicable diseases--Prevention. 4. Communicable diseases--Control. 5. Disease Vectors 6. Geographic Information Systems. 7. Geographic information systems--Mathematical models. 8. Spatial analysis (Statistics) I. Kumar, Lalit, 1960- II. Title.

RA643.K48 2015
614.4--dc23   2014046451

Visit the Taylor & Francis Web site at
http://www.taylorandfrancis.com

and the CRC Press Web site at
http://www.crcpress.com

Contents

List of Figures

Foreword

This book of modelling interactions between vector-borne diseases and the environment using geographic information system (GIS) methods fills many literature gaps. The book shows how GIS-based approaches provide innovative geographical methods with the capability of mapping and modelling such interactions with high accuracy. It shows how GISs can be used to merge satellite images with ground observations of vector demographics and disease incidence more accurately. It comes with the hope of increasing the ability of controlling the global prevalence of vector-borne diseases, such as dengue fever, malaria, and Rift Valley fever, which have increased dramatically in recent times, causing medical, environmental, and economic issues for most of the tropical and subtropical countries. Modelling interaction between vector-borne diseases and the environment using GISs increases understanding of the distribution of vector-borne disease incidence and vectors such as mosquitoes in time and space, which can be a major foundation for control and management programs for vector-borne diseases. The geographical methods used in this book show how knowledge of when and where disease cases and vectors occur enables the formulation of disease causation hypotheses for vectors and cases with unknown or poorly characterized aetiology, identification of areas at risk for disease, and design of efficient surveillance and control programs. These methods for modelling risks of diseases and vectors can also be implemented at local, country, and regional levels by vector-borne disease program managers, health officers and workers, and policy makers to ensure their optimal contribution to prevention, control, acceptance, and sustainability of programs. In addition, the book shows a variety of GIS implications in the planning of health interventions that can be used to enhance disease surveillance systems. It is useful for undergraduate and postgraduate students and postdoctoral researchers involved in epidemiological studies, particularly of vector-borne diseases, especially when they require the use of geographical modelling techniques in a GIS environment. The geographical modelling and analytical techniques described in this book are also valuable for researchers, workers, and students dealing with geographical data in the areas of entomology, environmental health, ecology, environmental science, public health, crime, geography, parasitological, and statistics. There is no doubt that GIS-based approaches will play a more significant role in such applications.

The first two chapters of the book give an overview of common vector-borne diseases, GIS-based mapping and modelling, impacts of climate change on vector distributions, and availability and importance of accurate epidemiologically relevant spatial data. They also describe modelling and simulating the prevalence of vector-borne diseases around the world and efforts directed toward combat and control, which would enrich the reader's knowledge about GIS applications on different types of vector-borne diseases. These chapters also summarize some key spatial techniques and how they can be used to aid in the analysis of geographical and attributed data. Beginning with Chapter 3, six chapters concentrate on the cartography and mapping of those diseases transmitted by mosquito-borne viruses, such as dengue fever and malaria, with examples of how such GIS-based mapping, in conjunction with its related tool-based approaches, can be used to visualize and analyse diseases and epidemiological data. Also, the chapters define the concept of establishing and characterizing spatial data systems, including their quality, errors, references and issues of scale, and building such a system from often quite separate, disparate sources. Simple means are used to describe common spatial methods for modelling and analysing spatial and temporal patterns and distributions of mosquito-borne diseases, spatial variation risk, and modelling associations between mosquito-borne diseases and environmental and socioeconomic factors, with illustrations of applications of the methods to a real data set collected from Saudi Arabia. The chapters show how to develop weather-based predictive modelling, which can be used to predict the weekly trend of vector abundance. It also shows a GIS case study for modelling the future potential distribution of vector-borne disease based on different climatic change scenarios. The predictive models described in this book would allow disease vector (e.g., mosquito) control measures to be initiated before sharp increases happen. Geographical analytic methods described throughout the book's chapters may be used to enhance vector-borne disease surveillance systems by better identifying and monitoring high-risk areas for disease and to find the association between disease distribution and different related factors.

I have no doubt that this book will be a welcome addition to the literature. It is a timely and useful contribution that emphasizes the significant role of GIS-based approaches in control of disease and public health problems.

Nabeel Koshak, PhD
Vice President for Business and Innovation,
Umm Al-Qura University
Vice Chair, BOD, Makkah Techno Valley Company
Director, King Abdulaziz City for Science and Technology
GIS Technology Innovation Centre

When my colleague Dr Hassan M. Khormi told me that he was writing a book with Associate Professor Lalit Kumar on modelling the interactions between vector-borne diseases and the environment using geographic information systems (GISs), I was delighted. The task seemed to me highly overwhelming.

It is an obvious multidimensional book, touching on medical geography, GIS, spatial analysis, and modelling. It so happens that I have taught all these subjects at the graduate and undergraduate college levels. I know well what it takes to carry out a huge task such as the one presented in this book. However, knowing the two authors well led me to expect the final quality of their work; we coauthored several papers together relevant to similar issues presented in this book.

It is evident that the book matches the latest state of the art in its multidisciplinary approach; it has transcended from the common single-facet approach, using traditional means, to more digitally automated measures for mapping and spatial analysis. Such a trend should appeal to a broad range of users and developers, especially younger generations and most of the high-tech enthusiasts in the areas of interest.

The authors embarked on engraving a piece of scientific art that satisfies a broad audience range: researchers, students, and decision makers from the health care arena, among many others. The expertise of both authors handy for tackling the profound elements of the book. The contents are not merely simple theoretical presentations of facts and methods on mapping and analysis, but are clearly documented and supported with actual research projects, undertaken by the two authors and others. Therefore, without any hesitation I would classify the present work as an evidence-based book, not on medicine, but on topics significantly relevant to medicine.

The diverse approach to the book from different perspectives has strengthened the various presentations and discussions; they include different aspects of relevant diseases, such as malaria, dengue fever, and Rift Valley fever. Furthermore, the flow of spatial representation, analysis, and discussion also varies; it includes microscale entities such as cities or even smaller sectors within cities. Then, it moves to higher-level spatial units, such as regions or continents. Finally, the impacts of climate change on vector-borne diseases are discussed at a global scale.

It is not common to find a similar book that would include such fruitful topics as in the present book: disease ecology and mapping, assessment of risk, and raster versus vector data representation and analysis. Several historical elements are also inserted in their appropriate positions in the text.

The technological advancements and the wisdom of the two authors have made the mission of this book easier and more universal in various ways. The authors come from two different parts of the world (Saudi Arabia and Australia). Despite physical barriers, they have succeeded in maintaining unique mentorship and collegial relationships. Therefore, they were able to solve the puzzles of vector-borne diseases in a clear scientific manner. As a result, we can become more aware and alert about why certain places are more susceptible to some vector-borne diseases, their prevalence, and their level of risk. It is definitely the hope that the scientific arguments of the book will add not only to literature but also to policy and action to eradicate vector-borne diseases.

It goes without saying that potential readers are well treated by the book's contents, regardless of any sophisticated mathematical or cartographical skills. I feel the book is inspiring and reassuring and adds significant value concerning automated mapping, GIS, and spatial analysis of relationships. Hence, it will have significant impacts on teaching and learning, research, and decision making.

From all of us who enjoy searching and researching on the broad complexity of disease ecology at large and digital disease mapping and analysis, thank you Hassan and Lalit. You have done a remarkable job. It is definitely worth celebrating and cherishing.

Ramze A. Elzahrany, PhD
Professor of Human Geography
Umm Al-Qura University
Makkah, Saudi Arabia

Preface

The geographic information system (GIS) has become an important tool for investigating, mapping, and managing vector-borne diseases. Many new techniques have evolved since the mid-2000s. In this book, we share our education and experience of over twenty years in the fields of GIS and vector-borne diseases. We have benefited from years of collaboration that have resulted in numerous journal publications and research reports. A book was a natural progression from papers and reports, especially because the field is one in which rapid advancements are being made, and there is easier access to spatial data. This book was inspired after having many publications, discussions, workshops, and work with vector-borne disease experts on GIS applications on modelling and mapping risks of the diseases. These discussions motivated us to put together the information acquired through years of research in a book so that scientists, students, GIS operators, and health workers can find information about the use of geographical techniques in vector-borne disease monitoring, management, and control. They also encouraged us to focus on step-by-step information for developing vector-borne disease risk models at different spatial and temporal scales.

We also aim to provide simple instructions to identify and visualize the hot spots and cold spots of disease incidence and vector abundance. This information is used to analyse the spatial patterns of vector-borne diseases, to model spatiotemporal risk changes in the incidence of vector-borne diseases, to assess vector-borne disease risks based on socioeconomic and environmental data, and to describe and analyse the relationship between the climatic variables and the disease vectors. Predictive models for estimating vector abundance and future potential distributions based on projected global climate change can be developed.

We use simple illustrative methods based on real data from dengue cases and the disease's vector, temperature, humidity, population and population density, and high-resolution satellite images. Using real data helped us introduce an innovative approach in applying GISs as a technology and applying spatial methods to study the spatial and temporal impacts of diseases and their vectors and their association with environmental and socioeconomic factors, in contrast to previous studies, which concentrated only on spatial aspects. Therefore, this book should help health-related communities better understand cutting-edge technologies, leading to better understanding of the vector-borne disease problem and its impacts on public health. We hope the techniques explained in this book inform decision makers in the field of vector-borne disease control of the importance of other factors related to the spread of the diseases and how they can be modelled to understand abundance and spread. The expected outcome of the book is to increase knowledge and enhance monitoring and control of the disease-caused infection, and we hope it will enable related disease control authorities to optimize high-quality control and surveillance systems.

Hassan M. Khormi
Lalit Kumar

Acknowledgements

We would like to thank King Abdulaziz City for Science and Technology (KACST) GIS Technology Innovation Centre at Umm Al-Qura University, Makkah, Saudi Arabia, for their support. We also thank Jeddah Health Affairs, Ministry of Health, KACST, Jeddah Municipality, the Central Department of Statistics and Information, and the Presidency of Meteorology and Environment for providing the data on dengue fever infection cases, satellite images, mosquito data, annual population data, and meteorological data used in some of the case studies. Thanks also go to Hanieh Saremi and Farzin Shabani for help with collation of information and to Veronica Cooper for helping with illustrations.

Many thanks to all our friends at the University of New England and at Umm Al-Qura University, especially those at the Department of Geography and KACST GIS Technology Innovation Centre, for their support and encouragement.

We would like to recognize the assistance rendered by CRC Press, Taylor & Francis Group, whose personnel supported this work and saw it to publication.

About the Authors

Dr Hassan M Khormi is assistant professor in the Department of Geography, Umm Al-Qura University (UQU), where he teaches about geographic information systems (GISs) and remote sensing. On completing his undergraduate degrees, he worked as a GIS operator in Jeddah Municipality from 2006 to 2008. In 2007, he joined UQU as a teaching assistant, after which he was awarded a scholarship to undertake master's and doctoral degrees in the field of GIS at the University of New England (UNE), Australia. The topic of his doctoral work was 'Modelling of Dengue Fever and Its Vector Risks Based on the Impacts of Socioeconomic, Meteorological and Environmental Factors: A Geographic Information System-Based Case Study of Jeddah, Saudi Arabia'. In early 2013, he moved back to Saudi Arabia to take a position as assistant professor in the Department of Geography and as deputy director of the GIS Technology Innovation Centre for administrative affairs. He is also a consultant (part-time) at Jeddah Municipality and an adjunct lecturer in the School of Environment and Rural Science, University of New England, Australia. His main research interests are in the fields of environmental modelling and GIS applications on vector-borne diseases.

Dr Lalit Kumar is an associate professor of spatial information technologies at the University of New England in Australia. He comes from Fiji, where he undertook his undergraduate and postgraduate studies in environmental science. He then moved to Sydney, where his doctoral studies were in the use of GISs and remote sensing for environmental modelling. After lecturing at the University of New South Wales in Sydney for a few years, he took up a position as assistant professor at the International Institute for Geo-Information Science and Earth Observation (ITC) in the Netherlands, where he stayed for five years. In 2002, he moved to the University of New England and has been there since that time.

Dr Kumar's expertise is in the use of GISs and remote-sensing technologies for mapping and modelling the environment, particularly natural resources and agricultural systems. He has undertaken extensive research on modelling the impacts of climate change on biodiversity, invasive species, and vector-borne diseases. He has undertaken research and consultancies in a number of countries, including Kenya, Tanzania, Ghana, Niger, Burkina Faso, India, Iran, Oman, Fiji, and Australia. He has over 150 publications, with over 100 journal articles in international peer-reviewed journals. Dr Kumar is also an associate editor for the journal *ISPRS Journal of Photogrammetry and Remote Sensing* and an academic editor for *PLoS ONE*.

Chapter 1

Introduction

1.1 Vector-Borne Diseases

According to the European Centre for Disease Prevention and Control (ECDPC, 2014), vector-borne diseases are 'infections transmitted by the bite of infected arthropod species, such as mosquitoes, ticks, triatomine bugs, sandflies, and blackflies'. Vector-borne diseases cause significant morbidity and mortality, especially in developing countries. They have accounted for large numbers of fatalities in humankind for a long period. In fact, up until the early twentieth century, vector-borne diseases accounted for more deaths in humans than all other causes combined (Kalluri et al., 2007). They were also responsible for the underdevelopment or nondevelopment of large areas of the tropics, especially in Africa. Vector-borne diseases account for seven of ten neglected infectious diseases that disproportionately affect poor and marginalized populations (Eisen and Eisen, 2011), two of the main ones being malaria and dengue fever. Malaria cases are estimated at around 250 million per year, with approximately 1 million deaths; dengue fever affects around 50 million people annually.

The World Health Organisation (WHO) describes disease vectors as any organism that is capable of transmitting disease parasites or pathogens from one infected host (human or animal) to humans (WHO, 2014). Vectors are generally arthropods (invertebrates). Arthropod vectors are especially sensitive to climatic factors as they are ectothermic (cold-blooded). Their rate of reproduction, survival, distribution and abundance, and habitability and their activities, such as biting, are influenced by weather (Lemon et al., 2008). They transmit pathogens of a viral, bacterial, or parasitical nature to humans or other warm-blooded hosts. The most common vectors are mosquitoes, ticks, triatomine bugs, sandflies, and blackflies.

Three key factors stand out when looking at vector-borne diseases: vector, space (location), and time. When a person becomes infected, there is a vector involved, and this determines the diseases that infect the individual. Then, there is the location, the position on Earth's surface where this event occurred. This position can be accurately located using a global positioning system (GPS) or estimated based on address. The location can be used to extract environmental conditions and meteorological variables for the position. Time of infection gives valuable information about when the vector is active. Together with location, time allows one to develop a spatiotemporal picture of the event; this aspect is extensively used in epidemiological modelling.

1.2 Mapping and Modelling Based on Geographic Information Systems

Mapping vector-borne diseases, like any other disease or spatial entity, is done by applying Tobler's first law of geography (Tobler, 1970): 'Everything is related to everything else, but near things are more related than distant things.' In other words, environmental and climatological conditions closer to a vector or disease occurrence should be more conducive to the vector's survival, reproduction, and transmission than the conditions further away. This suggests that study of the geographical location of pathogens and vectors, host interaction, environmental and climatological variables, and proximity to human or animal victims is paramount in understanding disease patterns. Spatial analysis answers questions such as what types of habitat contain the vectors, how far can vectors travel, which populations live in zones of high or low occurrences, and which other regions have similar conditions to where vectors are currently found so they may be denoted as high-risk areas.

The availability of georeferenced data, or spatial data, is of fundamental importance in mapping diseases and linking these to environmental risk factors. The geographical distribution and seasonal variation of vectors, vector abundance, and vector-borne disease transmission are predominantly controlled by environmental variables (such as land use, land cover, elevation, slope, aspect, and soil moisture) and climatological variables (such as temperature, rainfall, degree days, and humidity). Visual displays of quantitative data, such as cases of infection, on cartographic maps for the purpose of understanding causes have a long history. Perhaps the best-known case is that of Dr John Snow and the cholera deaths of London in the mid-nineteenth century (Snow, 1855). The visualization of spatial epidemiological data on an environmental or climatological background layer enables discerning patterns and correlations.

Early disease-mapping methods were mainly used for communicable diseases to identify sources of infection, rates of spread, and general environmental variables present at those sites (Howe, 1989). However, before computer-assisted mapping platforms were available, such mapping had severe limitations. It was generally only possible to have one layer of information at a time; any spatial analysis, even simple things like buffer and proximity analysis, needed to be undertaken manually, and it was difficult to handle large quantities of information. With the advent of the digital computer, particularly mainstream personal computers, geographic information systems (GISs) were developed, and quantitative thematic mapping started to flourish.

Generally, GISs are defined as computer-based systems for recording, storing, analysing, and displaying digital georeferenced data. They are regarded as a tool for linking and visualizing geographically referenced data, or spatial data, from different sources. However, this definition does not consider recent developments in which spatial analysis and modelling have become a core element of GISs. So, the GIS has moved from a storage-and-display platform to a decision support tool.

Current GIS-based epidemiological applications go far beyond the early studies of establishing correlates. The key aims of current disease mapping are to (1) describe the spatial variation in disease occurrence for formulating aetiological hypotheses; (2) identify clusters, hot spots, and areas of unusually high risk to formulate preventive action; (3) improve the reliability of disease risk maps for better resource allocation; and (4) undertake sophisticated spatial analysis of environmental factors and disease occurrence, together with geostatistical analysis, to statistically verify correlations (Rytkonen, 2004; Lawson et al., 1999, 2000). This fourth point is a new paradigm in epidemiological studies, and a number of chapters in this book are allocated to example applications. Modern GIS packages have strengthened the spatial statistics part so that quantitative analysis and hypothesis tests can be undertaken within the software package rather than exporting the data to specialist statistical packages. It is anticipated that with greater demand for robust quantitative modelling, the available statistical options within GIS packages will expand substantially in the future.

1.3 Impacts of Climate Change on Vector Distributions

Although biological and ecological determinants of vector-borne diseases are fairly well understood and researched, climate change impacts on future distribution of vectors are becoming an important area of epidemiological research. Reports have identified sharp upward trends over the next 50 to 100 years in global mean temperatures, sea level, ocean heat content, and frequency of extreme events, such as tropical cyclones (Church and White, 2006; Patz and Olson, 2008; Smith et al., 2014). These factors have the potential to directly influence vector-borne disease transmission by affecting the vector's geographic range. Increasing temperatures could increase rates of reproduction, affect the biting behaviour of vectors, shorten incubation periods of both pathogens and vectors, and increase the number of broods by providing longer suitable seasons or degree days. Regions or countries that are currently not a suitable habitat for the vectors may offer a conducive environment in the future, thus expanding the range of the vectors.

Climate change will also have an impact on rainfall patterns and thus on land cover. This in turn will change the microclimate and provide suitable environments for vectors as well as pathogens. Khormi and Kumar (2014) examined the potential additional risk posed by global climate change on the dengue vector *Aedes aegypti* distribution and abundance. Two global climate models were run with two emission scenarios, and distribution of the vector was modelled for 2030 and 2070. The results showed that many parts of the world that are currently not suitable for *Aedes aegypti* will become conducive for this vector to live and breed.

Ogden et al. (2008) reported an expansion in the range of *Ixodes scapularis*, thus leading to Lyme disease prevalence in central and eastern Canada. Risk maps based on climate change were developed using ambient air temperature, habitat, and number of ticks immigrating on migratory birds. Lemon et al. (2008) investigated the outbreaks of hantavirus pulmonary syndrome linked to specific climate events such as the El Niño Southern Oscillation (ENSO) index. It was suggested that the ENSO event of 1991–1992 created favourable conditions that led to an increase in the rodent population, resulting in the outbreak of hantavirus pulmonary syndrome.

Although these examples show longer-time climate impacts on vectors and vector-borne diseases, other studies have also looked at effects of seasonal variations on incidences. Patz et al. (1998) investigated the variability in biting rates of *Anopheles* mosquitoes with variability in weather patterns in Kenya. Large differences in reported cases of dengue fever in Jeddah, Saudi Arabia, were noted (Khormi and Kumar, 2012; Khormi, Kumar, and Elzahrany, 2011) based on changing climatological parameters, such as humidity, over different years.

1.4 Availability and Importance of Accurate Epidemiologically Relevant Spatial Data

The analysis and model outputs are only as good as the input data. This is not limited to epidemiological modelling but to any analysis that requires spatial data. However, this perhaps is of utmost importance in epidemiological modelling as decisions based on model outputs can have major impacts on large areas and populations. Spatial data used in modelling can have large associated errors, depending on scales and instruments used to capture such data. Modern GPSs have accuracies within ±5 m; however, if the data has been collected over long periods and date back a few decades, then the data could have accuracies of ±20 m or more. This does not mean that such data is not useful; in many instances, this is the only data that is available, and some data is better than none. However, the conclusions reached from the use of such data must clearly be used with relevant caveats.

The scale at which data is collected is also of paramount importance. Often, we see data collected at coarse scales being used in models that are then used to make predictions or decisions at much finer scales. Environmental and climatological factors determining the distribution and density of vectors operate differently at different scales, and it is important to determine the governing parameters operating at the scale of interest or the scales at which decisions are to be made. The models used should then have a higher weighting for these parameters, and outputs from these models are then valid for these scales only.

1.5 Structure of This Book

This book focuses on how advances in Geographic Information Systems and its inherent spatial analysis and modelling capabilities can be used to map, model, prevent, and control vector-borne diseases. It details some key spatial analysis techniques available in modern GISs, together with real-world applications of the techniques. The book is meant to show practitioners a sample of what is available and how these techniques could be used in the decision-making process.

Chapter 2 deals with modelling and simulating the prevalence of vector-borne diseases around the world and efforts on combat and control. It summarizes some key GIS-based epidemiological models and model types and discusses how they have been implemented. The chapter also summarizes some key spatial analysis techniques available in modern GIS packages and how they can be used to aid in the analysis of data.

Chapter 3 discusses cartographies and maps of vector-borne diseases. It highlights key elements that should be present in any map and how data generalizations affect outputs. It also discusses some key maps produced for dengue fever and malaria mapping. The chapter concludes by giving examples of different ways epidemiological data can be displayed on maps and includes point patterns, linear data, and areal displays (chloropleth and proportional).

Chapter 4 looks at spatial data, georeferencing, data types, data quality issues, accuracy, and scale issues. It also discusses data aggregation and its impact on model outputs. Remote sensing is a key data collection technology that can supply data over large areas at good spatial and temporal resolutions; it is increasingly being used in epidemiological studies. A discussion of remote-sensing data, capabilities, and potential uses is also included.

Chapter 5 incorporates a discussion of common spatial methods for modelling and analysing spatial and temporal patterns and distributions of mosquito-borne diseases. The discussion includes pattern analysis functions such as nearest neighbour, Getis-Ord Gi*, Ripley's K, and Moran's I. The distribution analysis discussion includes central features, standard deviation ellipse, linear directional mean, mean centre, and standard distance. The space and time pattern analysis functions of the Knox test and space-time K function are also discussed.

In Chapter 6, we discuss how the kernel and point density methods can be used to examine the level of spatial risk to identify and visualize where values of vector-borne disease incidences or the vector are geographically concentrated and homogeneous. We also describe how interpolation techniques such as kriging and inverse distance weighted (IDW) methods can be applied to generate spatially continuous surfaces.

Modelling associations between mosquito-borne diseases and environmental and socioeconomic factors is discussed in Chapter 7. It includes techniques such as geographically weighted regression, ordinary least squares, and Poisson, linear, and multiple regressions.

Chapter 8 discusses climate change impacts on potential future distributions of pathogens and vectors, together with modelled examples. A number of climate models, emission scenarios, and software such as CLIMEX are used to model potential future distribution of one vector, *A. aegypti*, as an example. The same techniques can be used for other vectors and vector-borne diseases.

The final chapter (Chapter 9) provides a conclusion and summarizes the contents of this book. It also suggests where research could head in the coming years.

References

Church, J.A., White, N.J. (2006). A 20th century acceleration in global sea level rise. *Geophysical Research Letters*, 33(1): L01602. doi:10.1029/2005GL024826

Eisen, L., Eisen, R.J. (2011). Using geographic information systems and decision support systems for the prediction, prevention, and control of vector-borne diseases. *Annual Review Entomology*, 56: 41–61.

European Centre for Disease Prevention and Control. (2014). Vector-borne diseases. Retrieved from http://www.ecdc.europa.eu/en/healthtopics/climate_change/health_effects/pages/vector_borne_diseases.aspx

Howe, G.M. (1989). Historical evolution of disease mapping in general and specifically of cancer mapping. *Recent Results Cancer Research*, 114: 1–21.

Kalluri, S., Gilruth, P., Rogers, D., Szczur, M. (2007). Surveillance of arthropod vector-borne infectious diseases using remote sensing techniques: a review. *PLoS Pathogens,* 3(10): e116.

Khormi, H., Kumar, L. (2014). Climate change and the potential global distribution of *Aedes aegypti*: spatial modelling using GIS and CLIMEX. *Geospatial Health,* 8(2): 405–415.

Khormi, H., Kumar, L. (2012). Assessing the risk for dengue fever based on socioeconomic and environmental variables in a GIS environment. *Geospatial Health*, 6(2): 171–176.

Khormi, H., Kumar, L., Elzahrany, R. (2011). Describing and analysing the association between meteorological variables and adult *Aedes Aegypti* mosquitoes. *Journal of Food, Agriculture and Environment*, 9: 954–959.

Lawson, A.B., Biggeri, A.B., Boehning, D., Lesaffre, E., Viel, J.F., Clark, A., Schlattmann, P., Divino, F. (2000). Disease mapping models: an empirical evaluation. Disease Mapping Collaborative Group. *Statistics in Medicine*, 19: 2217–2241.

Lawson, A.B., Böhning, D., Biggeri, A., Lesaffre, E., Viel, J.F. (1999). Disease mapping and its uses. In: Lawson, A., Biggeri, A., Böhning, D., Lesaffre, E., Viel, J.F., Bertollini, R. (Eds.), *Disease Mapping and Risk Assessment for Public Health*. West Sussex, UK: Wiley; 3–13.

Lemon, S.M., Sparling, P.F., Hamburg, M.A., Relman, D.A., Choffnes, E.R., Mack, A. (2008). *Vector Borne Diseases: Understanding the Environmental, Human Health and Ecological Connections*, Washington, DC: National Academies Press.

Ogden, N.H., Lindsay, L.R., Hanincová, K., Barker, I.K., Bigras-Poulin, M., Charron, D.F., Heagy, A., Francis, C.M., O'Callaghan, C.J., Schwartz, I., Thompson, R.A. (2008). The role of migratory birds in introduction and range expansion of *Ixodes scapularis* ticks, and *Borrelia burgdorferi* and *Anaplasma phagocytophilum* in Canada. *Applied Environmental Microbiology*, 74: 1780–1790.

Patz, J.A., Olson, S.H. (2008). Climate change and health: global to local influences on disease risk. In: Lemon, S.M., Sparling, P.F., Hamburg, M.A., Relman, D.A., Choffnes, E.R., Mack, A. (Eds.), *Vector-Borne Diseases: Understanding the Environmental, Human Health, and Ecological Connections*. Washington, DC: National Academies Press; 88–103.

Patz, J.A., Strzepek, K., Lele, S., Hedden, M., Greene, S., Noden, B., Hay, S.I., Kalkstein, L., Beier, J.C. (1998). Predicting key malaria transmission factors, biting and entomological inoculation rates, using modelled soil moisture in Kenya. *Tropical Medicine and International Health*, 3(10): 818–827.

Rytkonen, M.J.P. (2004). Not all maps are equal: GIS and spatial analysis in epidemiology. *International Journal of Circumpolar Health*, 63(1): 9–24.

Smith, K.R., Woodward, A., Campbell-Lendrum, D., Chadee, D.D., Honda, Y., Liu, Q., Olwoch, J.M., Revich, B., Sauerborn, R. (2014). Human health: impacts, adaptation, and co-benefits. In: Field, C.B., Barros, V.R., Dokken, D.J., Mach, K.J., Mastrandrea, M.D., Bilir, T.E., Chatterjee, M., Ebi, K.L., Estrada, Y.O., Genova, R.C., Girma, B., Kissel, E.S., Levy, A.N., MacCracken, S., Mastrandrea, P.R., White, L.L. (Eds.), *Climate Change 2014:*

Impacts, Adaptation, and Vulnerability. Part A: Global and Sectoral Aspects. Contribution of Working Group II to the Fifth Assessment Report of the Intergovernmental Panel on Climate Change. Cambridge: Cambridge University Press; 709–754.

Snow, J. (1855). *On the Mode of Communication of Cholera.* 2nd ed. London: Churchill.

Tobler, W. (1970). A computer movie simulating urban growth in the Detroit region. *Economic Geography*, 46(2): 234–240.

World Health Organisation (WHO). (2014). Vector-borne disease. Retrieved September 9, 2014, from the Health and Environment Linkages Initiative (HELI), http://www.who.int/heli/risks/vectors/vector/en/

Chapter 2

Modelling and Simulating the Prevalence of Vector-Borne Diseases around the World and Efforts for Combat and Control

2.1 Introduction

Vector-borne diseases pose a major threat to human populations, especially in the tropics, where populations of vectors such as mosquitoes flourish in abundance. Many of the most dangerous human diseases, such as malaria and dengue fever, are transmitted by insect vectors. The control of vector-borne diseases presents a major challenge to global health officials, as every year hundreds of millions of people suffer from them (World Health Organisation, 2014). Most of the approaches in the control of vector-borne diseases have focused on mathematical methods in describing the dynamics of the interactions of host and parasite or the simulation models in describing the parasite/vector behaviour in relation to environmental conditions (Gettinby et al., 1992). Previously, the control of these diseases was dependent on decreasing the number of infective bites either through the use of insecticide-treated bed nets or indoor residual spraying of insecticides (Coleman et al., 2006). However, the success of the insecticide-treated bed nets and indoor residual spraying programs are affected by the dual issues of parasites developing drug resistance and insecticide resistance in the vectors, and if they are not monitored directly, a significant increase in disease transmission can occur (Coleman et al., 2006). The resistance threatens the control system of the diseases as resistance evolves at a faster pace than newly developed antibiotics, drugs, and insecticides; therefore, sustainable resistance management is essential to effectively manage the vectors and vector-borne diseases.

A vector-borne disease is caused by an infectious microbe that is transmitted to a host via a blood-feeding arthropod (i.e., an insect or an arachnid). The most common vectors are mosquitoes and ticks, which carry diseases such as malaria, dengue fever, West Nile virus (WNV), and Lyme disease. The vector-borne diseases can be bacterial, viral, protozoan, or helminthic in nature. The vectors that carry the diseases have specific environments where they live and breed. These environments can be remotely sensed using satellites, and the conditions can be monitored (Kalluri et al., 2007). While eradicating these insects is next to impossible, the high-risk areas can be mapped and monitored using geographic information systems (GISs) and remote sensing.

The application modelling capabilities of GISs to vector-borne diseases has led to better and a more comprehensive understanding of variables that contribute to the spread of these diseases. The spatial analysis capabilities of modern GIS systems are incredibly vast and powerful, and simple processes can be combined and used as part of a more complex analysis. In relation to vector-borne diseases, spatial analysis can include landscape characterization, proximity of the disease to mapped features, cluster analysis, as well as a range of

environmental factors necessary for vector survival and reproduction. GIS systems are frequently used to identify and quantify the risk of disease occurrence, and many models have
been developed for these purposes. Other applications include
GIS-aided control techniques such as insecticides, as well as
developing low-cost monitoring systems that can be utilized
by poorer and developing communities. As the availability of
data increases and the quality and accuracy of remote-sensing
methods improve, the possible applications of GISs to a field
such as vector-borne diseases will continue to increase.

Over the years, technological advances, such as molecu-
lar techniques for pathogen and species identification as well
as rapid development in hardware and software capability
for data collection, analysis, and management, have emerged
in the battle against vector-borne diseases (Eisen and Eisen,
2011). Spatial information techniques, such as GISs, remote
sensing, and spatial statistics, have allowed researchers to
identify and model these disease patterns and examine their
relationships with environmental factors such as climate
(Kolivras, 2006). With better spatial statistics methods and
improved GIS capabilities, the spatial aspects of disease rates
and their transmission can be addressed more thoroughly
(Chaput et al., 2002).

Developing successful warning strategies to manage vector-
borne diseases can be difficult because of the dynamic nature
of the diseases, constantly changing environmental variables,
involved vectors and hosts, and health system infrastructure,
all of which need to be combined in an integrated manner (Racloz et al., 2012). However, many approaches have
been suggested recently for the surveillance of vector-borne
diseases, in particular integration approaches that combine
mathematical modelling and mapping with the use of spatial
analysis tools such as GISs. The use of such technologies not
only improves the monitoring, surveillance, and identification
of the control factors and determines present spatial patterns
of disease incidence in high-risk areas, but also helps in the

prediction of the disease outbreak and the occurrence of vector-borne pathogens (Pathirana et al., 2009; Winters et al., 2010). These spatial information techniques can be effective tools to fill the gaps in the current understanding of disease distribution (Pathirana et al., 2009).

2.2 GISs in Vector-Borne Disease Modelling

A GIS is a powerful technological tool that is capable of mapping and modelling vector-borne diseases and linking their prevalence to environmental and climatological factors. Vectors occur within quantifiable habitable zones that can be statistically correlated with geographical, climatic, and environmental factors. The strong spatial and temporal correlations shown by vector-borne diseases make them especially conducive to modelling and mapping using GISs. A variety of disease variables (vector distribution, presence, abundance, density, and disease incidence); environmental variables (land use type, soil type, distance from forests, urban areas, and elevation); and climatic variables (temperature, humidity, and precipitation) have been used to generate disease risk maps that aid in public health awareness, limited resource distribution, and prevention, control, and monitoring of vector-borne diseases.

The clustering behaviour of diseases can be mapped to generate spatial patterns. Hypothesis testing can then be undertaken by measuring correlations between the prevalence of vectors, diseases, and the surrounding spatial features or environmental conditions. The ability for GISs to store, visualise, and analyse data make them ideal for modelling spatial relationships. With the advent of high-resolution remote sensing, data sets such as those for rainfall, temperature, elevation, humidity, vegetation cover, water bodies, and population distribution and densities can be quickly and easily obtained and combined into a GIS model to predict the presence and abundance of vectors. A key benefit of GIS modelling is the ability to map target areas for prioritization of vector control (Sabel and Löytönen, 2004); this may be particularly important for regions with limited resources.

Spatial patterns of disease incidences are important for analysing and identifying their causes (Foody, 2006). One of the earliest, and perhaps the best-known example of the use of spatial data for health analysis purposes, was Dr John Snow's mapping of a cholera outbreak in Soho, London, in 1854 (Bynum, 2013). Snow meticulously mapped incidences of cholera in Soho and was able to correlate the case locations to a centrally located water pump, demonstrating that the contaminated water was responsible for the spread of the disease. Although cholera is not a vector-borne disease, this method of mapping is still applicable, and Snow created the awareness of how spatial data could be used in disease analysis. Spatial variables are especially useful in vector-borne disease modelling. Vector-borne diseases are known to be sensitive to environmental conditions because arthropod species are cold-blooded and require specific habitat conditions for survival and reproduction. Surrounding environmental and climatological conditions, such as humidity, vegetation type, temperature, proximity to and types of water source, are imperative to the survival of the host animal.

There are a number of ways in which GISs can be used to help control and mitigate vector-borne diseases. Generally, the spatial analysis of these diseases is undertaken by mapping the vectors themselves. The uses vary in the complexity and their applications. The use of GISs can be as simple as locating suitable sample sites or as complex as geostatistical modelling. For the most part, a GIS is used to create risk maps of vectors and diseases in an area of interest. Geostatistical modelling using GISs is an important method for generating risk maps. Statistical modelling allows interpretation of the environmental factors and predictors of risk. The risk maps produced can be used to inform the general public on vectors or diseases in a simple visual format. Risk maps can be used to direct disease control and detect-and-response emergency vector control, including priority area classification. They can be used to determine the dynamics of outbreaks by tracking the vector/disease path in the area. They can also be used in more complicated manners by helping to identify environmental and socioeconomic correlation to determine the risk of exposure (Eisen and Lozano-Fuentes, 2009).

The GIS analysis techniques used in vector-borne disease modelling include, but are not limited to, creating buffer zones around spatial features, calculating distances between spatial features; undertaking spatial clustering to identify incidence hot spots; building hydrological networks to calculate regions within a drainage basin that may be preferred vector breeding habitats, including slow-flowing streams or stagnant water; performing interpolation of data and statistical analysis to measure relationships in the spatial data; and explaining cause-and-effect relationships in the environment modelled. Once the causes or contributing factors are identified, models can be developed to identify regions that may be at risk of similar outbreaks. Identified outbreak clusters can be compared against other spatial data that is part of the GIS database. This could include a region's watershed, concentration of water bodies, microclimatological variables, elevation, land use, land cover, proximity to any type of flora/fauna, proximity to industrial processes, and so on.

Knowledge of the geographical distribution of vector-borne diseases is essential for understanding the global morbidity and mortality burdens and evaluating the impact of such activities (Bhatt et al., 2013). The growing use of georeferenced data and GISs has been transforming the analysis of the occurrence trend. In recent years, GISs have been used to visualize and identify the distribution of vector-borne diseases by spatial approaches, using household surveys, spatial point pattern analysis, and risk factor assessments, to determine the low- and high-prevalence areas (Khormi and Kumar, 2012). GISs have been used in a variety of research related to spatial patterns in vector-borne diseases, for example, in geographical distribution and gradients of the disease prevalence; identifying the populations at risk and geospatial and longitudinal disease trends; differentiating and defining risk factors; and in population health assessment, intervention planning, anticipating epidemics, and real-time monitoring of diseases (Kopp et al., 2002). A GIS-based method differs from traditional modes of disease reporting because it facilitates the standardization and integration of diverse data resources as well as analyses, retrieves, and manages data.

The GIS-based methods have been used to develop spatial models and visualise the patterns of risk for different vector-borne diseases in different parts of the world. For example, a database was linked with a GIS to design a surveillance system for displaying malaria risk in South Africa from 1995 to 1999 (Booman et al., 2000). The database enabled data on malaria incidences and their geographical locations (exact locations of villages and towns) to be recorded, providing a framework for designing control protocols. For this control program, digital map data sets of the entire region at different scales were collected, and three-dimensional (3D) displays of geographical areas (orthographs) were overlaid with aerial photographs. The mapping of malaria with the help of GISs allowed the investigation of the relationship between clinical cases of infection and the parasite exposure intensity and helped in the interpretation of the results by addressing some of the contributing factors, such as meteorological variables (Booman et al., 2000).

In another attempt, a malaria transmission intensity index (MTII) was constructed for Mexico through the investigation of demographic, socioeconomic, and ecological factors of malaria transmission using GIS information (Hernández-Avila et al., 2006). Information on the distribution of malaria cases was gathered for 12 years (from 1988 to 1999), and the MTII was created based on the number of cases and the duration and frequency of transmission outbreaks within the villages.

Several geographic and environmental variables, such as elevation, climate type, rainfall, evaporation, and vegetation type and coverage, were also included in this model. A logistic regression model was fitted to investigate the relationship between the local factors and the transmission of malaria. The estimated results of malaria incidence were plotted on the map, enabling easy determination of the areas of high incidence and defining the limits over time (Hernández-Avila et al., 2006).

Data sets can exhibit different spatial patterns at different spatial levels and scales (Mollié, 1999). Although GISs have been used to display the incidence of vector-borne diseases at a variety of administrative levels around the world (national and district levels), more studies are being undertaken at a microlevel and fine scale to show the nonuniformity of the disease within a district. However, the availability of organized spatial data at fine scales is still of importance (Rytkönen, 2004). The use of different systems and technologies has overcome some of the limitations. For example, the differential global positioning system (GPS) receivers have the ability to remove distortions caused by the signal travelling through the atmosphere; therefore, using them for collecting data allows the inaccuracies to be overcome and results in submetre accuracies (Martin et al., 2002).

Chaput et al. (2002) used GIS and spatial statistics, including techniques such as spatial filtering (smoothing) and cluster analysis, to display the spatial patterns of tick-borne disease in 1997–2000 in a 12-town area around Lyme, Connecticut. First, the home addresses of the people infected with the disease were mapped and geocoded to individual points using a GIS. Then, spatial filtering, which incorporates data from surrounding areas in an image or map, was undertaken. The disease rate was calculated using a minimum filter of 10 cases per area. If an area (census block group) did not meet the minimum 10 cases, then the smoothing method was performed around it. So, a circle was identified around the centroid of the census block group and was enlarged until it incorporated the centroid of the next-closest census block group. This process was repeated until the total cases circumscribed by the circle was greater than 10. The data was exported to the GIS from SAS software, and the annualized incidence by census block group was calculated based on the number of cases and larger population. The smoothing technique enabled the development of a clearer picture of the regions that were at increased risk on a finer geographic scale while maintaining the stability of the estimated disease rates. The use of the GIS and the spatial distribution of the maps in the study confirmed the cases and areas of high risk for tick-borne diseases (Chaput et al., 2002). These techniques provide an opportunity to investigate and quantify such diseases within an endemic area and therefore help in understanding the environmental factors contributing to vector-borne disease spread.

GISs have not only been used to visualize and model the distribution and risk patterns of different diseases in different parts of the world but also have been incorporated into information systems or decision support systems for the management of vector-borne diseases.

2.3 Decision Support Systems

Effective information management that can diagnose and control vector-borne diseases is based on three distinct phases: (1) data collection (from images, *in situ* measurements, and reports); (2) data analysis (information extraction and modelling); and (3) decision support (control purposes and predictive tools) (Kopp et al., 2002). There has been a concerted drive toward developing decision support systems to control vector-borne diseases. These interactive systems combine easy-to-use user-friendly data entry and data analysis capabilities to enable data presentation and model outcomes for decisions regarding allocation of surveillance and control resources (Eisen and Eisen, 2011). The decision support systems incorporate a variety of data from different resources, such as reports, disease incidences, vector surveillance records, disease control intervention monitoring, and stock control. These systems also use the GIS as a reporting tool to produce a variety of maps, graphs, and tables (Eisen and Eisen, 2011). The key benefits of these systems include the increased and improved capacity for electronic data storage, the collection of enormous amounts of data in a single system, and the improved capacity for monitoring the control program performance (Eisen and Eisen, 2011). These systems can also be used without licensing costs and can be implemented in resource-poor environments.

Three questions need to be answered when developing an effective decision support system. First, what determines the acceptable level of disease risk? This usually depends on the goals of the public health officials of a region or a country. Second, is the quantification of the required vector density designed to reduce or eliminate disease transmission? This requires information about the relationships between vector and parasite transmission (e.g., modelling between different factors and vector population). Third, what is the best way to measure entomological risk? These three questions can form a conceptual basis that addresses the disease-specific differences as well as control strategies (Hemingway et al., 2006).

A number of different programs have been suggested to improve the tools and technologies available for vector-borne diseases. An example of this is the Innovative Vector Control Consortium (IVCC), a program funded by the Bill and Melinda Gates Foundation established in 2005 (Hemingway et al., 2006). This program is designed to reduce transmission of mosquito-borne diseases around the home with innovative tools. The IVCC has two main objectives: to develop improved insecticides and to provide better tools for a decision support system at the community level. The IVCC has developed an information tool that includes data, decision support software, and online access to central databases of entomology and epidemiology within a GIS-based format (Hemingway et al., 2006).

Another example is the GIS-based malaria information system (MIS), which was developed for the malarious provinces of South Africa to provide relevant information for decision making and research (Martin et al., 2002). The MIS, which is a system based on the personal computer (PC), consists of a user-friendly front end for data input; an automated mapping component running off a vector-based GIS; a relational database (malaria cases, population, home addresses, etc.); spatial map data, including boundaries, roads, and health facilities; and a user guide. The MIS also includes the weekly summary data regarding the spraying of the targeted areas with insecticide (Martin et al., 2002). The resulting maps from the MIS showed the incidence of the malaria in the region, which supported formulation of malaria insecticide and drug policies, helped control malaria, and provided appropriate information on malaria transmission and a decision support platform for regional malaria control.

Although disease mapping has long been used in medical geography and epidemiology, with the main aims to describe the spatial variation in disease incidence, identify areas of high risk for preventive action, and provide reliable maps of disease risk in the region, the development of advanced spatial analysis tools in GISs has resulted in more complex spatial analysis of disease incidences and the contributions of environmental variables.

2.4 Spatial Analysis Capabilities of GISs

As mentioned, GISs can access different information from a wide variety of sources and link and visualize data geographically. For example, GPS data points and aerial or satellite images can be linked to a GIS for easy integration of information. GISs can eliminate the duplication of effort involved in data collection and therefore reduce cost. Surveillance and monitoring of vector-borne diseases requires continuous collection and analysis of data (Abdullahi, 2013). The spatial scale of analysis or the level of precision can be chosen in the GIS, with smaller areas and scales of research interest providing more specific findings.

The display of disease incidence data can range from simple point maps and graphics for disease cases to the mapping of potential risk from complex disease models (Lawson et al., 1999). Disease mapping is essentially a risk estimation of disease potential across a chosen study area and is usually used for health resource allocation. The pattern of description and the distribution of diseases can be presented in dot, diagram, choropleth, and flow maps, all of which can be achieved relatively easily through the use of a GIS.

In addition to the abilities of GISs in data collection, display, and presentation, much of the strength of GIS is derived from its spatial analysis capabilities. Spatial analysis is referred to as 'the ability to manipulate spatial data into different forms and extract additional meaning as a result' (Bailey, 1994, p. 15). It involves a range of methods and procedures, such as geography and statistics for analysing the data (Clarke et al., 1996). The spatial analysis task includes visualization, exploratory data analysis, and model building (Gatrell and Bailey, 1996) and varies in complexity depending on the task (e.g., simple overlay analysis to statistical models) (Clarke et al., 1996). The overlay analysis can combine the characteristics and compute new values of several data sets into one layer. The sensitivity of the results to the weights and cutoff values in the layers can be determined by the GIS computational and visual display capabilities (Clarke et al., 1996).

A GIS can also create buffer zones (proximity analysis) around selected features and areas that meet the designated criteria. The user can specify the buffer size depending on the purpose and combine the information with disease incidence to identify the exact number of cases within the buffer. The ability to create a buffer zone around hazardous sites, polluted locations, water bodies, and so on can be useful in health research and vector control strategies.

A GIS can undertake specific calculations, such as those for distance, areas, and proportion of population within a certain feature or radius. The statistical analysis in a GIS can also be used to explore data, in particular when defining classes and ranges on a map; summarize data; and identify and confirm spatial patterns. Public health professionals are increasingly using GISs (adding spatial data to GISs, producing maps and interpreting the results and outcomes by the patterns shown on the map) (Rytkönen, 2004). Employing linear and multiple regression analysis to identify contributions of various factors in disease spread and abundance is also used for understanding the spatial variation in disease prevalence.

Visualization can be used in novel ways to interpret and explore the results of traditional statistical analysis and can show the changes in disease patterns over time (Clarke et al., 1996). Three-dimensional GIS animation analysis (ArcGlobe and ArcScene) is also considered an effective way to show the spread or retreat of diseases over space and time.

The exploratory spatial data analysis tools available in modern GISs provide insights about data distribution, clustering, and outliers; levels of spatial autocorrelation; and variation among multiple datasets (Esri, 2013). Advances in computing and display technology have made exploratory spatial analysis one of the most active areas in GIS/spatial analysis research (Clarke et al., 1996). Disease clustering evaluates spatial epidemiological data to identify aggregated spatial and space-time patterns of disease occurrence.

One of the exploratory methods most used to identify disease incidence and determine the presence of local clustering of census tracts is the use of hot spot analysis based on the Getis-Ord Gi* statistic. It identifies statistically significant hot spots and cold spots (local level of spatial cluster) within a set of weighted features within a context of neighbouring features to identify regions affected by the disease (Khormi and Kumar, 2011a; Mitchell, 2012).

Other exploratory spatial and temporal data algorithms, which are used to determine patterns in data and to measure the spatial dependency with its neighbours as specified by sample data, include geographically weighted Poisson regression (GWPR), local indicators of spatial association (LISA) statistics, multilogistic regression, local Moran's I index, and Geary's Index (Khormi and Kumar, 2011a). These algorithms tend to find patterns of features and attributes that occur in space and time with distinction to local clusters or local spatial outliers (Gajović and Todorović, 2013). For example, spatial autocorrelation can be measured with Moran's I, values ranging from +1 as perfect correlation and −1 as perfect dispersion (Moran, 1950).

Ecological analysis examines the spatial relationship between the vector or disease and environmental and climatological covariates in the region of study. Such spatial analysis methods can be used to model disease risk as a function of one or more environmental or climatological variables, essentially modelling the impact environmental and climatological factors have on arthropod vectors and their transmission of pathogens to host reservoirs. In general, modelling involves the integration of GIS data layers with standard statistical methods. For example, the risk of dengue fever prevalence was modelled based on various socioeconomic parameters, nationality, and age groups using GISs and remote sensing in Saudi Arabia (Khormi and Kumar, 2011b). The method used for this purpose involved detailed neighbourhood quality information (e.g., roof area of houses and street widths). Geographically weighted regression (GWR) was also used for analysing socioeconomic parameters, such as population density, and a prediction model was created based on the dengue fever cases and the related socioeconomic factors. The results indicated high positive association between the disease and socioeconomic factors; also, dengue fever was more prevalent in adults between the ages of 16 and 60 (Khormi and Kumar, 2011b).

Spatial autocorrelation is also an important statistical analysis method in the development of vector abundance risk models. Spatial autocorrelation takes into account the spatial heterogeneity inherent in the distribution of vector-borne diseases and therefore reduces biased estimates of the coefficients and their standard errors and affects model predictions (Legendre, 1993). Wimberly et al. (2008) reported that the incorporation of either spatial autocorrelation (the tendency for pathogen distributions to be clustered in space) or spatial heterogeneity (the potential for environmental relationships to vary spatially) has improved the accuracy of the predictive disease risk models and the environmental predictions of the geographic distributions of pathogens and vectors.

Other GIS models, such as spatial interaction and spatial diffusion models, have also been used in disease-related studies. Spatial interaction models are able to analyse and predict the movements of people and hosts from place to place and therefore identify areas with high risk of disease transmission (effects of distance and population size) (Clarke et al., 1996). Spatial diffusion models are similar to spatial interaction models; however, they incorporate time and space along with basic epidemiologic concepts; therefore, they predict and analyse the spread of the disease spatially and temporally from infected to susceptible people in a region (Thomas, 1990).

All of the examples suggest GISs as a powerful and innovative tool and an important component that can be used for studies of health and control of environmental diseases. The speed of data manipulation in a GIS and its ability to handle large amounts of data, undertake repetitive tasks, handle map projections and scales, create buffer zones, and provide detailed cartographic output are some of the advantages of a GIS. The independence of a GIS and its ability to simultaneously use text, sound, animation, and graphics have made this system highly desirable for modelling.

2.5 Examples of GIS-Based Modelling of Vector-Borne Diseases

The GIS has been used for modelling a number of vector-borne diseases and in a wide variety of disease models, both for mapping and for prediction of risk maps and likely future distributions. This section provides a small sample of applications in some key vector-borne diseases. This is not an exhaustive list of diseases or application types; however, it does give the breadth of applications and scenarios to show the applicability of spatial analysis techniques for disease modelling.

2.5.1 Malaria

Mosquitoes are the most common carriers of vector-borne diseases, and malaria is the most severe disease that they can carry. Malaria is often fatal as successful treatment needs to occur as soon as the patient shows symptoms. Malaria cases are generally in clusters as the environments for the infected mosquitoes are localized. Coleman et al. (2009) studied malaria cases in Mpumalanga Province, South Africa, as they occurred between 2002 and 2005. SaTScan software was used to map spatial and temporal clusters of malaria. There were 422 cases identified in 341 households across 7 towns during the 3 main malaria seasons. The research reported a high degree of spatial clustering, with the spatial clusters changing locations across the years. In the 2004/2005 malaria season, the use of GIS-based software helped to direct malaria control efforts because of this spatial and temporal clustering.

Yang et al. (2010) mapped the spread of malaria throughout China. Using GISs, they combined rainfall, relative humidity, and temperature information with statistical modelling to determine the increasing rate of days per year that mosquitoes could reproduce, based on previous 15 years of climate data. A risk map was produced that showed the changes in potential days that mosquitoes could breed and hence the increased potential risk.

Ra et al. (2012) applied a multiple linear regression model to remotely sense and provide GIS data to identify risk of malaria in Varanasi District, India. Variables included precipitation, temperature, Normalized Difference Vegetation Index (NDVI), elevation, land cover, land use, population density, and distance to features such as health care, roads, ponds, and water sources. The study reported a strong correlation between distance from health care facilities and incidences of malaria.

Another study that utilized regression modelling with environmental and climatological variables was that by Jeganathan et al. (2001). They looked at two mosquito species, *Anopheles dirus* and *Anopheles minimus*, in the neighbouring regions of North Lakhimpur and Dibrugarh, North East India, to quantify breeding habitats of the vectors in two different landscapes sharing similar climatic conditions. Extensive field surveying was used to determine vector distribution; land cover type was determined through remote sensing. Buffer zones of 1 km were applied to each land cover type to include the potential distances mosquitoes are known to travel from breeding sites, and then each land cover type was correlated with vector distribution data. *Anopheles dirus* was found to occur in association with the forest fringes; *Anopheles minimus* was found to occur within close proximity to slow-flowing streams shaded with vegetation.

Daash et al. (2009) used the GIS with a view to identify risk factors for a malaria outbreak based on various ecological parameters in the Koraput District, Orissa, India. Data layers used included recorded outbreaks, geomorphology, soil type, drainage networks, land use, natural water bodies (including streams), forest areas, and human settlements. Results of the modelling exercise indicated that the regions in the study area host to the highest malaria incidences contained high forest cover, extensive stream networks, and a large number of valleys. Regions with the lowest incidence of malaria transmission contained less forest cover and good drainage networks.

Nmor et al. (2013) applied logistic regression modelling to a number of topographic variables to determine the variables more likely to correlate with the location of malaria breeding sites. Data used included those from the Shuttle Radar Topography Mission 3 (SRTM3) and Advanced Spaceborne Thermal Emission Reflection Radiometer (ASTER) digital elevation models (DEMs). The modelling results enabled the characterisation of topographic features more likely to correlate with the location of malaria vector breeding sites and the prediction of the locations of previously undetected sites.

Schröder et al. (2007) used GIS-based modelling to determine the likelihood of disease proliferation within a potential host population in the event that a single infected individual was introduced into a colony. The variables used in the model included relative density of females in the colony, proportion of blood meals taken from humans versus animals, and average ambient temperatures. Static factors used in the calculations included upper and lower parasite growth threshold temperatures, timing of breeding patterns and gonotrophic cycles, vector survival rates, and the rate of human recovery from infection. The results of the modelling showed that a present-day malaria outbreak in certain parts of Lower Saxony, Germany, could lead to a transmission period of up to 3 months. Summer months were deemed to be the highest risk, with almost the whole country a potential transmission zone. The study also tested the transmission model using climate-warming scenarios for 2020, 2060, and 2100 predicted by the Intergovernmental Panel on Climate Change (IPCC); it determined that the maximum risk of a 4-month transmission period would be in 2100, with the highest-risk period lasting from May through October.

de Oliveira et al. (2013) used GIS analysis and logistic regression for the analysis and identification of the relative likelihood and social determinants of malaria infection in the Vale do Amanhecer rural settlement in Brazil. Remote sensing and GIS data were integrated to predict disease likelihood so that it could be used for future planning. Spatial data used included land use and land cover, vegetation density, terrain slope and downslope direction, and distance of domiciles from possible breeding grounds and gold-mining sites described in the de Oliveira et al. (2013) study. The study reported high cases of malaria in places with a high population density and in areas with selective logging and mining activities. Similarly, Palaniyandi (2014) and Palaniyandi and Mariappan (2013) used logistic regression techniques to develop maps to help control the spread of diseases such as dengue, malaria, and chikungunya throughout India. Because of its fertile land, India has a large mosquito population that provides vectors for these diseases. GIS technology was used to map environmental determinants of mosquito abundance, track disease distribution, and identify hot spots for management.

2.5.2 Dengue Fever

Porcasi et al. (2012) used a GIS to design a risk stratification model for dengue fever at national and urban levels in Argentina. The model required input layers of vector populations and densities, present and historical cases, population densities, and current mosquito control activity. The environmental, disease, and control layers were added together and then multiplied by a weighted factor to determine the total risk for each location.

Chang et al. (2009) developed a GIS surveillance model for Nicaragua by utilizing satellite imagery and data from local epidemiologic surveys to create a cost-effective method of identifying risk so that resources could be more focused on controlling the virus. Neighbourhood boundaries were digitized, and risk factors for each neighbourhood were entered into an attribute table, including access, water availability, and mosquito infestation incidences. Other spatial layers that were used included stagnant water sources and mosquito larvae presence data collected through surveys.

We (Khormi and Kumar, 2012) created dengue risk maps for Jeddah, Saudi Arabia, using a number of environmental, climatological, socioeconomic, and disease incidence layers. One of the main aims of the risk maps was to support resource allocation in the control of the vector *Aedes aegypti*. The variables used in the model included population density, mosquito counts, confirmed dengue cases, water access, and neighbourhood quality information. A total of 111 districts were investigated; 15% were identified as having a high risk of dengue fever, 22% had medium risk, and the remainder had low or very low risk. The authors suggested that the model could be implemented as a routine procedure for the control of the dengue vector and thus prevention of dengue fever.

When they developed a model for Tainan City, Taiwan, Chu et al. (2013) also had the idea of improved efficiency and resource allocation of insecticide spraying to control the dengue vector *Aedes aegypti* mosquito. The data used for the analysis included a base map of the region, dengue surveillance data, and insecticide spraying records. Risk levels were assessed based on frequency, intensity, and duration of dengue outbreaks. The model was useful in ensuring adequate coverage was provided by the insecticide spraying program, as well as in calculating the most efficient use in developing regions where resources are limited. By mapping where and when dengue had occurred, GIS data analysis was able to show the frequency, duration, and intensity of dengue outbreaks throughout the year to model where targeted insecticide control should be distributed. The key idea was to shift from insecticide spraying as a reactive process to one that was proactive. The model was able to optimize the spraying area to only 21% of the historical spraying area, creating a far more efficient spraying program. The model and application showed that a GIS-aided control process can contribute to a vector management framework by providing an integrated approach and enabling evidence-based decision making on insecticide spraying patterns.

Kolivras (2006) developed static maps of suitable dengue vector mosquito habitat in Hawaii; GIS software and such variables such as precipitation, temperature, and stream or wetland locations, together with areas populated by humans, were used to determine the locations most likely to experience dengue transmission if the virus were introduced. The maps developed presented a useful tool to agencies and other stakeholders responsible for controlling dengue and its vectors.

2.5.3 Rift Valley Fever

Rift Valley fever, common in western Africa and some parts of the Middle East, is spread by the mosquito *Aedes vexans*. These mosquitoes breed in stagnant water holes, where they lay eggs on the side of the holes. The eggs then need at least 10 mm of rainfall, followed by 6 dry days and then more rainfall. The virus in the mosquitoes can survive in unhatched eggs. Risk zones for Rift Valley fever can be established by constructing zones of potential mosquitoes using spatial data for rainfall and locations of the appropriate water holes. Marechal et al. (2008) developed such risk maps of Rift Valley fever for the Ferlo region of Senegal in West Africa using high-resolution images from the SPOT-5 satellite, which was used to derive a Normalized Difference Pond Index (NDPI) that showed the water bodies. Turbidity of these ponds was calculated using the Normalized Difference Turbidity Index (NDTI), which was then used to determine all those water bodies that were suitable for *Aedes vexans* breeding. Different risk zones were established as very high risk (0–50 m), high risk (50–150 m), and moderate risk (150–500 m) from the suitable pond sites.

Vignolles et al. (2009) expanded on this risk map by adding rainfall data, both frequency and intensity, to determine if the eggs would hatch. They also used high-spatial-resolution Quickbird imagery to identify areas where animals were fenced in, along with parks and villages in the area. This data was factored into the risk map to improve the forecasting ability.

2.5.4 *Visceral Leishmaniasis (Kala-Azar)*

Visceral leishmaniasis (also known as *kala-azar*) is a vector-borne disease caused by the parasitic protozoa of the genus *Leishmania*. It is transmitted to humans and other hosts through the bites of sandflies, which breed in forest areas, caves, or the burrows of small rodents.

On the Indian subcontinent, the bulk of kala-azar (as leishmaniasis is known here) occurs in the northeastern regions near the Himalayan mountains and Bangladesh, specifically in the regions of Uttar Pradesh, Jharkhand, Bihar, and West Bengal. Bhunia et al. (2010) used GIS-based modelling to determine the geographical distribution of leishmaniasis in northeastern India. They used disease incidence, topography, vegetation, climatological data, and health and census information along with the socioeconomic status of the inflicted population. Results showed that infection rates increased with increasing population density and higher temperature and rainfall. The areas that had the lowest literacy rates, the lowest income, and the lowest socioeconomic status were shown to have the highest incidences of leishmaniasis.

Rossi et al. (2007) investigated the effect of environmental differences, such as land use, on the distribution densities of sandflies around Mt Vesuvius. Samples from the coastal and the Apennine side of the volcano were collected, together with land use, elevation, slope, and aspect information. The results showed that there was a statistically significant increase in sandflies in the agricultural and urban areas compared to natural forested areas. However, there was no significant difference between each side of the mountain or any significant difference between human or canine cases of leishmaniasis on either side of the mountain.

2.5.5 Lyme Disease

Lyme disease (*Lyme borreliosis*) is the most common tick-borne disease amongst people in the Northern Hemisphere. It is caused by at least three species of bacteria belonging to the genus *Borrelia* and transmitted to humans via three different species of ticks (Kalluri et al., 2007). In Europe, the primary cause is the sheep tick, usually the females in the nymphal stage of their life cycle.

Tack et al. (2012) determined the influence of vegetation cover, specifically shrub, herb, and forest cover, on tick abundance in Antwerp and Limburg in northeastern Belgium. They sampled 21 different forest sites in the region for ticks and added these locations to a digital forest cover map using GIS software. Three-kilometre buffers were created around these locations to determine the landscape and vegetation cover for the immediate area. The samples were then taken and the captured ticks were tested for the bacteria that results in Lyme disease. They concluded that an estimated mean of 9.1% of captured ticks were carrying the bacteria at these locations. While forest, herb, and shrub cover showed little difference in the results, there was a significant increase in tick population amongst the oak trees.

Using GIS software, Leighton et al. (2012) modelled time-to-establishment rates for *Ixodes scapularis* tick populations (a primary vector of the Lyme disease pathogen *Borrelia burgdorferi* in North America) based on rainfall, temperature, soil composition, elevation, and vapour pressure. The spatial distribution and prevalence of white-tailed deer (*Odocoileus virginianus* Zimmermann), which is a key host for adult *I. scapularis*, along with the locations of deciduous woodland and coniferous forest regions were also taken into account. The movement and migration behaviours of deer and passerine birds were also included because they are a means by which ticks can disperse and travel great distances to other potential host populations. The results of the modelling exercise indicated temperature was the most important factor in infection development; this suggested that future climate warming may facilitate further expansion in the range of Lyme disease risk zones in North America.

A second study on deer tick distributions was conducted by Guerra et al. (2002), with Wisconsin, northern Illinois, and northern Michigan the key US areas of interest. Sites were selected from forests and parks across the three states, and small mammals collected at these sample sites were checked for ticks. Tick samples were also collected by dragging a flannel cloth through the vegetation and recording how many ticks were captured per hour. Soil samples for each forest type location were added as layers to a GIS. Climate, environmental, geological, and rainfall data was used in the statistical analysis. The results of the modelling showed that only forest type and soil type were significant in the prevalence of ticks at the sample sites.

2.5.6 Chagas Disease (American Trypanosomiasis)

Chagas disease (*American trypanosomiasis*) is a tropical parasitic disease transmitted to humans by the protozoan *Trypanosoma cruzi* on being bitten by the insect *Triatominae* (kissing bugs). At the moment, the disease is endemic to Central and South America. Lambert et al. (2008) produced risk maps showing the extent of the US population that was at risk during each month of the triatomine breeding season.

The minimum temperature for triatomine reproduction is 19°C, so data was collected for the average minimum temperature in the United States for the months of June, July, and August. These were overlaid with US population densities, together with known cases of Chagas disease and locations of three species of triatomine. Modelling showed that Texas was most at risk, with cases already reported and the majority of the state in the risk area in all 3 months. Florida, Louisiana, southern California, and southern Arizona were also at risk during these 3 months. All of the other states in the southern United States were at risk but to a lesser extent.

Barbu et al. (2013) developed a GIS-based model to understand the dynamics and distribution of Chagas disease and its spread through the vector *Triatoma infestans* in Arequipa, Peru. Their study measured patterns in the distribution of the disease vector between neighbouring households, houses across the street from one another, the presence or absence of animals, and the presence of particular building materials. They concluded that over 90% of infestation was able to be correlated with infected neighbours within the same block and that roads were an effective barrier for vector spread. The modelling enabled a better understanding of disease dynamics, an important step in combating and controlling the spread of the disease.

In Africa, Chagas disease has been shown to be spread by tsetse flies. Control of tsetse populations could lead to larger areas of land being opened to livestock or mixed crop-livestock enterprises; however, a poorly planned eradication program could result in accelerated bush clearing, reduction of vegetative cover, and increased runoff and erosion (Symeonakis et al., 2007). Symeonakis et al. combined remotely sensed and other environmental and ancillary data in a GIS-based decision support system to help inform tsetse control programs and limit possible detrimental side effects. Data layers used included distribution of tsetse, land designation, richness in bird species, cattle density, crop use intensity, and erosion risk. The results of their modelling enabled clear differentiation and identification of priority areas for tsetse control under a range of hypothetical scenarios.

Other tsetse-related studies that have used GIS as a modelling tool to map the fly's distribution include those of Rogers and Williams (1993, 1994) in Zimbabwe; Rogers and Randolph (1993) and Rogers and Williams (1993, 1994) in Kenya and Tanzania; Rogers et al. (1996), Hendrickx (1999), and Hendrickx et al. (2001) in West Africa; and Robinson et al. (1997a, 1997b) in Southern Africa.

2.5.7 West Nile Virus

West Nile virus is a mosquito-borne arbovirus typically found in temperate and tropical regions of the world. It was first identified in the West Nile subregion in the East African nation of Uganda in 1937 but has now spread globally. WNV is considered to be an endemic pathogen in Africa, the Middle East, Asia, Europe, Australia, and the United States. Mosquitoes are the prime vector for the transmission of WNV, with birds the most commonly infected animal and serving as the prime reservoir host. WNV has been found in various species of ticks, but current research suggests they are not important vectors of the virus.

Young et al. (2013) investigated the environmental factors contributing to the distribution of WNV in the United States. Environmental variables used in the modelling included precipitation, temperature, NDVI, elevation, and land cover. The study was able to achieve a correlation of up to 0.86 between predicted values and actual values, suggesting that remotely sensed environmental data could be used to predict future risk and incidence of WNV. However, there was significant variation in the model's output for different time periods, suggesting that the relationship between contributing environmental factors varied from year to year.

Zou et al. (2007) developed a spatial analysis tool to estimate the potential WNV distribution based on degree-day estimation of the ambient temperatures exceeding the lower threshold value needed for effective virus incubation. The model calculated whether the extrinsic incubation period (EIP) for the virus would be completed within the average longevity of an individual *Culex tarsalis* Coquillett mosquito vector. The resulting model could be used successfully to delineate areas that would potentially be host to WNV activity and thus had the potential to be used to direct personnel and funding into areas of high risk as predicted by the model.

A number of studies have recently utilized data from the avian population to identify risk levels of WNV. Although modelling of bird population exposure to vectors such as mosquitoes is generally seen as too difficult because of large and sparse distributions, dead bird counts have been identified as an indicator that human cases will also be occurring (Cooke et al., 2006). Cooke et al. used dead bird reports alongside other environmental factors to improve the risk-modelling process in Mississippi. A number of environmental variables were used with dead bird counts; however, each variable was weighted based on the variable significance determined using statistical probability. Environmental factors that were considered relevant were slope, road density, stream density, vegetation, aspect, soil permeability, and climatic variables. Data for human and avian cases of WNV in Mississippi was collected and separated according to zip codes. The model was able to identify a number of the environmental variables that were related to higher levels of WNV risk. The model also showed that WNV risk was correlated with WNV occurrence in avian populations; thus, bird-based risk maps could be an accurate indicator of environmental conditions suitable for sustaining the virus. Similar modelling approaches have also recently been applied to WNV in Greece, using data from wild birds, equids, and chickens (Valiakos et al., 2014).

2.5.8 Other Vector-Borne Diseases

A study undertaken by Vescio et al. (2012) investigated the relationship between a number of environmental factors, such as mean temperature; NDVI; mean habitat fragmentation level; proportion of areas covered by grasslands, scrub, and herbaceous vegetation; cattle, sheep, and human density; and the incidence of Crimean-Congo haemorrhagic fever in Bulgaria. The model outputs included incident rate ratios, odds ratios, and the 95% confidence levels for each of the environmental factors. The study was able to identify a number of environmental factors that correlated with higher rates of infection, leading the researchers to conclude that the findings could be useful in supporting disease prevention initiatives in a number of ways, such as risk identification and encouraging behavioural change under high-risk conditions.

The prevalence of Buruli ulcer disease along the Densu River in southern Ghana was studied by Kenu et al. (2014) to determine the spatial distribution of the disease throughout the region and quantify the relative risk to the local population living within the catchment. Variables such as disease incidence, water body locations, pollution levels, fields, and roads were used to derive the predictive model. Clustering analysis showed that the disease tended to occur in areas where the river had slowed sufficiently to become a habitat for vectors and where rivers showed the highest signs of pollution.

Vector distribution, land cover, elevation data, and the presence of host species were used by Jacob et al. (2010) to assess areas most at risk for the vector-borne viral disease eastern equine encephalomyelitis in Tuskegee, Alabama. The study specifically focused on the vector mosquito species, *Culex erraticus,* and host bird species, northern cardinal. Extensive field surveys mapped the distribution of mosquito species and passerine birds known to be a vector of the disease. The study combined land cover maps derived from Quickbird satellite images with georeferenced mosquito- and bird-sampling sites to evaluate corresponding land cover attributes. Kriging techniques were used to interpolate total mosquito and bird counts and develop a spatial linear prediction model for locating potential mosquito and bird habitats. The GIS-based modelling was able to identify the regions that were most habitable for both vector and host species.

Andreo et al. (2014) used a species distribution modelling (SDM) approach in a GIS to model the statistical correlation between the location of hantavirus pulmonary syndrome infections, caused by Andes virus, and remotely sensed climatic and environmental conditions to predict the areas in southern Argentina where people were at the highest risk of contracting the disease. The vector for this disease is the rodent *Oligoryzomys longicaudatus*. Disease incidence data from 1995 to 2009 and 19 climatic variables and 5 classes of land cover information were used in a GIS-based model. The areas determined to be most at risk were those with high rainfall, dry summers, cold winters, and little bare soils and dominated by subantarctic forest and shrubland. The research also reported that the areas in which infection risk was highest overlapped only 28% with the most suitable habitat for *O. longicaudatus*. Andreo et al. (2014) proposed the use of models such as that developed in the study as a tool for public health authorities to focus efforts and resources more efficiently.

Bluetongue, an arboviral disease that affects certain breeds of sheep and is spread via the bites of *Culicoides* midges, was the subject of a study undertaken by Ducheyne et al. (2007). *Culicoides* actively migrate using the wind, with flights as long as 100 km in high wind being reported (Ducheyne et al., 2007). This study used the migration pattern of the insects from an outbreak in Greece and Bulgaria between 1999 and 2001 and a GIS to model potential new infection. Wind direction and speed, temperature, rainfall, and georeferenced outbreak data were used to show that it was the wind more than any other method of migration that is used by the *Culicoides*.

Canine dirofilariasis is a mosquito-borne disease that affects dogs and other carnivores and is endemic in many European countries, including Italy. Mortarino et al. (2008) developed a model with the objective of detecting new cases, confirming prevalence, evaluating risk factors, characterising environmental factors, and estimating the yearly number of infections of *Canine dirofilariasis*. Disease incidence, together with other data layers, such as administrative boundaries, elevation, temperature, rainfall, and humidity, were used in a GIS-based model. The results of the research showed that there was a higher probability of infection for hunting and truffle dogs compared to guard or pet dogs. No relationship was found for breed, sex, and habit.

Every year around 2.7 billion passengers are transported by airlines around the world (Huang et al., 2012), and this airline network and its rapid expansion increases the risk of connecting regions of endemic vector-borne disease with the rest of the world. This in turn increases the challenges faced by nations in terms of health systems to deal with vector-borne importation into the country and its invasion events. Huang et al. (2012) developed a web-based GIS tool, known as the Vector-Borne Disease Airline Importation Risk Tool (VBD-AIR), to help understand the roles of airports and airlines in the transmission and spread of vector-borne diseases. Vectors can travel in the aircraft to international destinations within moderately high atmospheric pressure space (Huang et al., 2012). However, the role of air travel in moving the vector-borne disease is much more significant than moving the vector itself (Tatem et al., 2012). It is reported that 8% of travellers to developing countries become ill on returning home, and this is mostly from vector-borne diseases (Huang et al., 2012). Data assembled consisted of the spatial data sets on modelled global disease and vector distributions, along with the climatic and air network traffic data. The model can be used by decision makers to examine the risk factors for disease spread and develop surveillance and control methods for dealing with the risks more efficiently.

A few representative examples of the use of GISs in vector-borne disease modelling were mentioned. A search of the literature shows many more examples and for many more diseases. A search of the literature also indicates that there have been many more studies in the last decade compared to previous periods; this is mainly because of better capabilities in the area of spatial analysis and modelling in modern GIS packages as well as greater availability of high-resolution spatial data, mainly from recently launched satellites.

2.6 Conclusion

Vector-borne diseases place large financial burdens on public health systems, especially in developing countries. With no drug and vaccination to control some of these diseases, the key solution is the surveillance and monitoring of these vectors and the spatial distribution of the disease (Khormi and Kumar, 2012). Maps are fundamental tools for detecting the spatial patterns of the diseases and vectors. By using GISs, it is possible to link data of research interest and the geographically referenced information according to map coordinates (Rytkönen, 2004). However, the GIS offers much more than simple mapping because of its extensive data management, analysis, and display capabilities. Much of the value of a GIS is because of its spatial analysis and modelling capability. GISs can reveal spatial variations and distributional patterns of diseases. With the assistance of the disease maps, low- and high-risk areas can be highlighted and environmental factors related to the disease can be determined; consequently, the causes behind the prevalence of the disease can be defined (Rytkönen, 2004; Khormi and Kumar, 2011a). Thus, GISs have become a major player in the prediction and prevention of vector-borne diseases. The application of GIS models to map, monitor, and develop control routines for vector-borne diseases around the world has contributed to a far greater understanding of the conditions under which they spread and has aided the implementation of more efficient control measures.

Although the GIS system can provide high levels of insight into spatial patterns, there are a number of potential issues associated with the use of GISs for disease mapping and health monitoring. One of these challenges derives from the lag experienced between contact with the disease and the onset of symptoms. This spatiotemporal element can provide misleading results because of movements of the population between the time of infection and the onset of symptoms (Foody, 2006). Data quality is another major issue in the use of GISs for these applications (Sipe and Dale, 2003). Data quality issues include inconsistent and incomplete disease-reporting methods, which in many instances vary between districts within a country. Difficulty in merging this data compromises the analysis that can be performed. Despite these drawbacks, the benefits that come from the application of GISs to vector-borne disease mapping and modelling far outweigh the limitations. The models are continuously improving our ability to identify risk factors and allow for preemptive control measures to be employed. As data becomes more accessible, the quality of the data improves, more powerful spatial analysis techniques are included in GIS software, and models continue to develop, the application of GIS in epidemiology will continue to provide greater benefits and more efficient solutions.

References

Abdullahi, F.B. (2013). A comparative analysis of geographical information system (GIS) and electronic health (E-Health) in decision making. *International Journal of Research in Engineering, IT and Social Sciences,* 3: 61–71.

Andreo, V., Neteler, M., Rocchini, D., Provensal, C., Levis, S., Porcasi, X., Rizzoli, A., Lanfri, M., Scavuzzo, M., Pini, N., Enria, D., Polop, J. (2014). Estimating hantavirus risk in Southern Argentina: a GIS-based approach combining human cases and host distribution. *Viruses,* 6: 201–222.

Bailey, T. (1994). A review of statistical spatial analysis in geographical information systems. In Fotheringham, S., Rogerson, P. (Eds.), *Spatial Analysis and GIS*. London: Taylor and Francis.

Barbu, C.M., Hong, A., Manne, J.M., Small, D.S., Quintanilla Calderón, J.E., Sethuraman, K., Quispe-Machaca, V., Ancca-Juárez, J., Cornejo del Carpio, J.G., Málaga Chavez, F.S., Náquira, C., Levy, M.Z. (2013). The effects of city streets on an urban disease vector. *PLoS Computational Biology*, 9(1): e1002801. doi:10.1371/journal.pcbi.100280.

Bhatt, S., Gething, P.W., Brady, O.J., Messina, J.P., Farlow, A.W., Moyes, C.L., Drake, J.M., Brownstein, J.S., Hoen, A.G., Sankoh, O., Myers, M.F., George, D.B., Jaenisch, T., Wint, G.R.W., Simmons, C.P., Scott, T.W., Farrar, J.J., Hay, S.I. (2013). The global distribution and burden of dengue. *Nature*, 496: 504–507.

Bhunia, G.S., Kesari, S., Jeyaram, A., Kumar, V., Das, P. (2010). Influence of topography on the endemicity of Kala-azar: a study based on remote sensing and geographical information. *Geospatial Health*, 4(2): 155–165.

Booman, M., Durrheim, D.N., La Grange, K., Martin, C., Mabuza, A.M., Zitha, A., Mbokazi, F.M., Fraser, C., Sharp, B.L. (2000). Using a geographical information system to plan a malaria control programme in South Africa. *Bulletin of the World Health Organization*, 78: 1438–1444.

Bynum, W.F. (2013). On the mode of communication of cholera: W.F. Bynum reassesses the work of John Snow, the Victorian 'cholera cartographer'. *Nature*, 495(7440): 169–170.

Chang, A.Y., Parrales, M. E., Jimenez, J., Sobieszczyk, M.E., Hammer, S.M., Copenhaver, D.J., Kulkarni, R.P. (2009). Combining Google Earth and GIS mapping technologies in a dengue surveillance system for developing countries. *International Journal of Health Geographics,* 8: 49. doi:10.1186/1476-072X-8-49

Chaput, E.K., Meek, J.I., Heimer, R. (2002). Spatial analysis of human granulocytic ehrlichiosis near Lyme, Connecticut. *Emerging Infectious Diseases,* 8: 943–948.

Chu, H.J., Chan, T.C., Jao, F.J. (2013). GIS-aided planning of insecticide spraying to control dengue transmission. *International Journal of Health Geographics*, 12: 42. doi:10.1186/1476-072X-12-42

Clarke, K.C., McLafferty, S.L., Tempalski, B.J. (1996). On epidemiology and geographic information systems: a review and discussion of future directions. *Emerging Infectious Diseases*, 2: 85–92.

Coleman, M., Coleman, M., Mabuza, A.M., Kok, G., Coetzee, M., Durrheim, D.N. (2009). Using the SaTScan method to detect local malaria clusters for guiding malaria control programmes. *Malaria Journal*, 8: 68. doi:10.1186/1475-2875-8-68

Coleman, M., Sharp, B., Seocharan, I., Hemingway, J. (2006). Developing an evidence-based decision support system for rational insecticide choice in the control of African malaria vectors. *Journal of Medical Entomology*, 43: 663–668.

Cooke, W.H., III, Grala, K., Wallis, R.C. (2006). Avian GIS models signal human risk for West Nile virus in Mississippi. *International Journal of Health Geographics*, 5: 36. doi:10.1186/1476-072X-5-36

Daash, A., Srivastava, A., Nagpal, B.N., Saxena, R. Gupta, S.K. (2009). Geographical information system (GIS) in decision support to control malaria—a case study of Koraput district in Orissa, India. *Journal of Vector Borne Diseases*, 46: 72–74.

de Oliveira, E.D., dos Santos, E.S., Zeilhofer, P., Souza-Santos, R., Atanaka-Santos, M. (2013). Geographic information systems and logistic regression for high-resolution malaria risk mapping in a rural settlement of the southern Brazilian Amazon. *Malaria Journal*, 12(1): 420. doi:10.1186/1475-2875-12-420

Ducheyne, E., De Deken, R., Bécu, S., Codina, B., Nomikou, K., Mangana-Vougiaki, O., Georgiev, G., Purse, B.V., Hendrickx, G. (2007). Quantifying the wind dispersal of *Culicoides* species in Greece and Bulgaria. *Geospatial Health*, 1(2): 177–189.

Eisen, L., Eisen, R.J. (2011). Using geographic information systems and decision support systems for the prediction, prevention, and control of vector-borne diseases. *Annual Review of Entomology*, 56: 41–61.

Eisen, L., Lozano-Fuentes, S. (2009). Use of mapping and spatial and space-time modeling approaches in operational control of *Aedes aegypti* and dengue. *PLoS Neglected Tropical Diseases*, 3(4): e411. doi:10.1371/journal.pntd.0000411

Esri. (2013). ArcMap 10.2. Redlands, CA: Esri.

Foody, G.M. (2006). GIS: health applications. *Progress in Physical Geography*, 30(5): 691–695.

Gajović, V., Todorović, B. (2013). Spatial and temporal analysis of fires in Serbia for period 2000-2013. *Journal of the Geographical Institute Jovan Cvijic, SASA,* 63: 297–312.

Gatrell, A.C., Bailey, T.C. (1996). Interactive spatial data analysis in medical geography. *Social Science and Medicine,* 42: 843–855.

Gettinby, G., Revie, C., Forsyth, A.J. (1992). Modelling: a review of systems and approaches for vector-transmitted and other parasitic diseases in developing countries. Proceedings of a workshop organised jointly by the International Laboratory for Research on Animal Diseases and the Food and Agriculture Organisation of the United Nations, Nairobi, Kenya.

Guerra, M., Walker, E., Jones, C., Paskewitz, S., Cortinas, M.R., Stancil, A., Beck, L., Bobo, M., Kitron, U. (2002). Predicting the risk of Lyme disease: habitat suitability for *Ixodes scapularis* in the North Central United States. *Emerging Infectious Diseases,* 8(3): 289–297.

Hemingway, J., Beaty, B.J., Rowland, M., Scott, T.W., Sharp, B.L. (2006). The innovative vector control consortium: improved control of mosquito-borne diseases. *Trends in Parasitology,* 22: 308–312.

Hendrickx, G. (1999). Georeferenced Decision Support Methodology Towards Trypanosomosis Management in West Africa. PhD thesis, University of Gent.

Hendrickx, G., de La Rocque, S., Reid, R., Wint, W. (2001). Spatial trypanosomosis management: from data-layers to decision making. *Trends in Parasitology,* 17: 35–41.

Hernández-Avila, J.E., Rodríguez, M.H., Betanzos-Reyes, A.F., Danis-Lozano, R., Méndez-Galván, J.F., Velázquez-Monroy, O.J., Tapia-Conyer, R. (2006). Determinant factors for malaria transmission on the coast of Oaxaca State, the main residual transmission focus in Mexico. *Salud Pública de México,* 48: 405–417.

Huang, Z., Das, A., Qiu, Y., Tatem, A.J. (2012). Web-based GIS: the vector-borne disease airline importation risk (VBD-AIR) tool. *International Journal of Health Geographics,* 11: 1–14.

Jacob, B.G., Burkett-Cadena, N.D., Luvall, J.C., Parcak, S.H., McClure, C.J.W., Estep, L.K., Hill, G.E., Cupp, E.W., Novak, R.J., Unnasch, T.R. (2010). Developing GIS-based eastern equine encephalitis vector-host models in Tuskegee, Alabama. *International Journal of Health Geographics,* 9: 12. doi:10.1186/1476-072X-9-12

Jeganathan, C., Khan, S.A., Chandra, R., Singh, H., Srivastava, V., Raju, P.L.N. (2001). Characterisation of malaria vector habitats using remote sensing and GIS. *Journal of the Indian Society of Remote Sensing,* 29(1): 31–36.

Kalluri, S., Gilruth, P., Rogers, D., Szczur, M. (2007). Surveillance of arthropod vector-borne infectious diseases using remote sensing techniques: a review. *PLoS Pathogens,* 3(10): e116. doi:0.1371/journal.ppat.0030116

Kenu, E., Ganu, V., Calys-Tagoe, B.N.L., Yiran, G.A.B., Lartey, M., Adanu, R. (2014). Application of geographical information system (GIS) technology in the control of Buruli ulcer in Ghana. *BMC Public Health,* 14: 724. doi:10.1186/1471-2458-14-724

Khormi, H.M., Kumar, L. (2011a). Identifying and visualizing spatial patterns and hot spots of clinically-confirmed dengue fever cases and female *Aedes aegypti* mosquitoes in Jeddah, Saudi Arabia. *Dengue Bulletin,* 35: 15–34.

Khormi, H.M., Kumar, L. (2011b). Modeling dengue fever risk based on socioeconomic parameters, nationality and age groups: GIS and remote sensing based case study. *Science of the Total Environment,* 409: 4713–4719.

Khormi, H.M., Kumar, L. (2012). Assessing the risk for dengue fever based on socioeconomic and environmental variables in a geographical information system environment. *Geospatial Health,* 6: 171–176.

Kolivras, K.N. (2006). Mosquito habitat and dengue risk potential in Hawaii: a conceptual framework and GIS application. *The Professional Geographer,* 58(2): 139–154.

Kopp, S., Shuchman, R., Strecher, V., Gueye, M., Ledlow, J., Philip, T., Grodzinski, A. (2002). Chapter 3: public health applications. *Telemedicine Journal and e-Health,* 8: 35–48.

Lambert, R.C., Kolivras, K.N., Resler, L.M., Brewster, C.C., Paulson, S.L. (2008). The potential for emergence of Chagas disease in the United States. *Geospatial Health,* 2(2): 227–239.

Lawson, A.B., Böhning, D., Biggeri, A., Lesaffre, E., Viel, J.F. (1999). Disease mapping and its uses. In: Lawson, A., Biggeri, A., Böhning, D., Lesaffre, E., Viel, J.F., Bertollini, R. (Eds.), *Disease Mapping and Risk Assessment for Public Health.* Chichester, UK: Wiley; 3–13.

Legendre, P. (1993). Spatial autocorrelation—trouble or new paradigm? *Ecology,* 74: 1659–1673.

Leighton, P.A., Koffi, J.K., Pelcat, Y., Lindsay, L.R., Ogden, N.H. (2012). Predicting the speed of tick invasion: an empirical model of range expansion for the Lyme disease vector *Ixodes scapularis* in Canada. *Journal of Applied Ecology*, 49: 457–464.

Marechal, F., Ribeiro, N., Lafaye, M., Güell, A. (2008). Satellite imaging and vector-borne diseases: the approach of the French National Space Agency (CNES). *Geospatial Health*, 3(1): 1–5.

Martin, C., Curtis, B., Fraser, C., Sharp, B. (2002). The use of a GIS-based malaria information system for malaria research and control in South Africa. *Health and Place*, 8: 227–236.

Mitchell, A. (2012). *The Esri Guide to GIS Analysis*. Redlands, CA: Esri Press.

Mollié, A. (1999). Bayesian and empirical Bayes approaches to disease mapping. In: Lawson, A., Biggeri, A., Böhning, D., Lesaffre, E., Viel, J.F., Bertollini, R. (Eds.), *Disease Mapping and Risk Assessment for Public Health*. Chichester, UK: Wiley; 15–29.

Moran, P.A.P. (1950). Notes on continuous stochastic phenomena. *Biometrika*, 37: 17–23.

Mortarino, M., Musella, V., Costa, V., Genchi, C., Cringoli, G., Rinaldi, L. (2008). GIS modeling for canine dirofilariosis risk assessment in central Italy. *Geospatial Health*, 2(2): 253–261.

Nmor, J.C., Sunahara, T., Goto, K., Futami, K., Sonye, G., Akweywa, P., Dida, G., Minakawa, N. (2013). Topographic models for predicting malaria vector breeding habitats: potential tools for vector control managers. *Parasites and Vectors*, 6: 14. doi:10.1186/1756-3305-6-14

Palaniyandi, M. (2014). GIS for disease surveillance and health information management in India. *Geospatial Today*, 13(5): 44–46.

Palaniyandi, M., Mariappan, T. (2013) Containing the vector borne diseases, (GIS for master plan for mosquito control in the metropolitan cities in India). *Journal of Geospatial Today*, 12(8): 28–30.

Pathirana, S., Kawabata, M., Goonetilake, R. (2009). Study of potential risk of dengue disease outbreak in Sri Lanka using GIS and statistical modelling. *Journal of Rural and Tropical Public Health*, 8: 8–17.

Porcasi, X., Rotela, C.H., Introini, M.V., Frutos, N., Lanfri, S., Peralta, G., De Elia, E.A., Lanfri, M.A., Scavuzzo, C.M. (2012). An operative dengue risk stratification system in Argentina based on geospatial technology. *Geospatial Health*, 6(3): S31–S42.

Ra, P.K., Nathawat, M.S., Onagh, M. (2012). Application of multiple linear regression model through GIS and remote sensing for malaria mapping in Varanasi District, India. *Health Science Journal,* 6(4): 731–749.

Racloz, V., Ramsey, R., Tong, S., Hu, W. (2012). Surveillance of dengue fever virus: a review of epidemiological models and early warning systems. *PLoS Neglected Tropical Diseases,* 6(5): e1648. doi:10.1371/journal.pntd.0001648

Robinson, T., Rogers, D.J., Williams, B. (1997a). Mapping tsetse habitat suitability in the common fly belt of southern Africa using multivariate analysis of climate and remotely sensed vegetation data. *Medical and Veterinary Entomology,* 11: 235–245.

Robinson, T.P., Rogers, D.J., Williams, B. (1997b). Univariate analysis of tsetse habitat in the common fly belt of southern Africa using climate and remotely sensed vegetation data. *Medical and Veterinary Entomology,* 11: 223–234.

Rogers, D.J., Hay, S.I., Packer, M.J. (1996). Predicting the distribution of tsetse flies in West Africa using temporal Fourier processed meteorological satellite data. *Annals of Tropical Medicine and Parasitology,* 90: 225–241.

Rogers, D.J., Randolph, S.E. (1993). Distribution of tsetse and ticks in Africa: past, present, and future. *Parasitology Today,* 9: 266–271.

Rogers, D.J., Williams, B.G. (1993). Monitoring trypanosomiasis in space and time. *Parasitology,* 106: 77–92.

Rogers, D.J., Williams, B.G. (1994). Tsetse distribution in Africa: seeing the wood and the trees. In: Edwards, P.J., May, R.M., Webb, N. (Eds.), *Large-Scale Ecology and Conservation Biology.* Oxford, UK: Blackwell Scientific; 247–271.

Rossi, E., Rinaldi, L., Musella, V., Veneziano, V., Carbone, S., Gradoni, L., Cringoli, G., Maroli, M. (2007). Mapping the main *Leishmania phlebotomine* vector in the endemic focus of the Mt Vesuvius in southern Italy. *Geospatial Health,* 2: 191–198.

Rytkönen, M. (2004). Not all maps are equal: GIS and spatial analysis in epidemiology. *International Journal of Circumpolar Health,* 63: 9–24.

Sabel, C.E., Löytönen, M. (2004). Clustering of disease. In: Maheswaran, R., Craglia, M. (Eds.), *GIS in Public Health Practice.* London: CRC Press; 51–67.

Schröder, W., Schmidt, G., Bast, H., Pesch, R. Kiel, E. (2007). Pilot-study on GIS-based risk modelling of a climate warming induced tertian malaria outbreak in Lower Saxony (Germany). *Environmental Monitoring and Assessment*, 133: 483–493.

Sipe, N.G., Dale, P. (2003) Challenges in using geographic information systems (GIS) to understand and control malaria in Indonesia. *Malaria Journal*, 2: 36. doi:10.1186/1475-2875-2-36

Symeonakis, E., Robinson, T., Drake, N. (2007). GIS and multiple-criteria evaluation for the optimisation of tsetse fly eradication programmes. *Environmental Monitoring and Assessment*, 124: 89–103.

Tack, W., Madder, M., Baeten, L., Vanhellemont, M., Gruwez, R., Verheyen, K. (2012). Local habitat and landscape affect *Ixodes ricinus* tick abundances in forests on poor, sandy soils. *Forest Ecology Management*, 265: 30–36.

Tatem, A.J., Huang, Z., Das, A., Qi, Q., Roth, J., Qiu, Y., (2012). Air travel and vector-borne disease movement. *Parasitology*, 139(14): 1816–1830. doi:10.1017/S0031182012000352

Thomas, R. (1990). *Geomedical Systems: Intervention and Control.* New York: Routledge; 314.

Valiakos, G., Papaspyropoulos, K., Giannakopoulos, A., Birtsas, P., Tsiodras, S., Hutchings, M.R., Spyrou, V., Pervanidou, D., Athanasiou, L.V., Papadopoulos, N., Tsokana, C., Baka, A., Manolakou, K., Chatzopoulos, D., Artois, M., Yon, L., Hannant, D., Petrovska, L., Hadjichristodoulou, C., Billinis, C. (2014). Use of wild bird surveillance, human case data and GIS spatial analysis for predicting spatial distribution of West Nile Virus in Greece. *Plos One*, 9(5): e96935. doi:10.1371/journal.pone.0096935

Vescio, F.M., Busani, L., Mughini-Gras, L., Khoury, C., Avellis, L., Taseva, E., Rezza, G., Christova, I. (2012). Environmental correlates of Crimean-Congo haemorrhagic fever incidence in Bulgaria. *BMC Public Health*. 12: 1116. doi:10.1186/1471-2458-12-1116

Vignolles, C., Lacaux, J.-P., Tourre, Y.M., Bigeard, G., Ndione, J.-A., Lafaye, M. (2009). Rift Valley fever in a zone potentially occupied by *Aedes vexans* in Senegal: dynamics and risk mapping. *Geospatial Health*, 3(2): 211–220.

Wimberly, M., Baer, A., Yabsley, M. (2008). Enhanced spatial models for predicting the geographic distributions of tick-borne pathogens. *International Journal of Health Geographics,* 7: 15. doi:10.1186/1476-072X-7-15

Winters, A.M., Eisen, R.J., Delorey, M.J., Fischer, M., Nasci, R.S., Zielinski-Gutierrez, E., Moore, C.G., Pape, W.J., Eisen, L. (2010). Spatial risk assessments based on vector-borne disease epidemiologic data: importance of scale for West Nile virus disease in Colorado. *American Journal of Tropical Medicine and Hygiene,* 82: 945–953.

World Health Organisation (WHO). (2014). *Policy Brief: IVM—The Power of Integrated Health and Environment Action.* Geneva, Switzerland: WHO.

Yang, G.-J., Gao, Q., Zhou, S.-S., Malone, J.B., McCarroll, J.C., Tanner, M., Vounatsou, P., Bergquist, R., Utzinger, J., Zhou, X.-N. (2010). Mapping and predicting malaria transmission in the People's Republic of China, using integrated biology-driven and statistical models. *Geospatial Health*, 5(1): 11–22.

Young, S.G. Tullis, J.A. Cothren, J. (2013). A remote sensing and GIS assisted landscape epidemiology approach to West Nile virus. *Applied Geography.* 45: 241–249.

Zou, L., Miller, S.N., Schmidtmann, E.T. (2007). A GIS tool to estimate West Nile Virus risk based on a degree-day model. *Environmental Monitoring and Assessment*, 129: 413–420.

Chapter 3

Cartographies and Maps of Vector-Borne Diseases

3.1 Introduction

Cartography and the mapping of vector-borne diseases are organized by their relativity to each other. Their strengths are derived from their basic approaches, with location being a principal attribute of the characterization of events. They are also a graphic way of thinking (Eisen and Lozano-Fuentes, 2009). Mapping diseases furthermore invites associations to the following bidirectionality, for example: If these disease cases are clustered around that lake, then that lake is central to these cases.

Medical science is also playing catchup with vector-borne disease issues. Its advances are a response to those diseases that we foster through social, economic, and environmental choices (Hakre et al., 2004). The cartography and mapping of disease blossomed in the nineteenth century during a period of vastly increasing international trade and emigration and again during the new era of globalization (Robinson, 1982).

In recent times, cartography and the mapping of vector-borne diseases using a geographic information system (GIS) has become firmly established. New concepts of cartography and the mapping of disease in a GIS environment continue to transform our ability to gather, analyse, and plot this disease data. A GIS for disease maps might serve as a metaphor for those social and environmental conditions that occupy geographic space wherein we may often feel inclined to look for emerging spatial solutions to diseases. Such analysis can shed light on a population and those locations at which thinking about improvements in public health can occur. This is essentially for the purpose of exploring disease issues and discovering ways to address them; it has taken its place in the conceptual and methodological foundations of vector-borne diseases (Khormi and Kumar, 2011). Mapping vector-borne disease distributions and hot spots and their associations with other environmental, climatic, and socioeconomic conditions is one of the best ways of helping forecast, control, monitor, and facilitate early detection of such conditions and aid in their prevention. Mapping diseases using a GIS has also shown promising results in assessing the risks of various vector-borne diseases such as dengue fever (DF) and malaria at different spatial scales (Snow et al., 1998).

Although a GIS and its related techniques cannot identify the vectors of the diseases themselves, it can characterize the environment in which such vectors thrive. As a new surveillance tool, it is also a powerful predictor in the mapping and modelling of the geographical limits, intensity, and dynamics of the risk of disease (Shirayama et al., 2009). This chapter concentrates on the cartography and mapping of those diseases transmitted by mosquito-borne viruses, such as dengue fever and malaria. In addition, in this chapter, we narrate with examples how such a GIS-based mapping, in conjunction with its related tools-based approaches, can be used to visualise and analyse diseases and epidemiological data and to describe those factors that can help in the control of diseases. Using a GIS and other methods in conjunction with climatic, socioeconomic, and environmental factors and mosquito distribution patterns, moreover, it should be possible to identify the risk areas at a predetermined spatial scale of investigation.

3.2 Cartographies of Vector-Borne Diseases

In general, cartographies of diseases such as vector-borne diseases involve the study and practice of making maps or charts for such diseases. These build on the promise that reality can be modelled in ways that communicate locational information effectively through combining sciences, aesthetics, and other technologies. In essence, mapmaking can be understood as a way of summarizing information pertaining to vector-borne diseases such as dengue fever in a graphic or practical way, for example, by mapping their nature, diffusion, or the way *Aedes aegypti* (as the main vector of the illness) is spread through a neighbourhood or other areas (Khormi and Kumar, 2012).

When developing maps, it is important to consider what format they should take. For example, the process of developing a map to be used in a printed book or journal will be different from developing a map to be displayed on a web page. Specifically, the use of colour and detailed symbols to identify cartographic features can be incorporated to display a map on a computer screen; more restricted use of black and white or a simpler symbology may be suitable for print media, such as journal articles.

However, any map must show basic details with key information concerning the mapped data that can be readily interpreted by readers (Figure 3.1). Such details might include the map's title (Figure 3.1a), scale (Figure 3.1c and 3.1e), legend (Figure 3.1d), features, direction symbol(s) (north) (Figure 3.1b), neat lines, temporal data of production, projection (Figure 3.1g), data sources, and agency (Figure 3.1f). Almost all maps incorporate some of these elements, such as scale, direction symbols, legend(s), and data source; other elements, such as neat lines, locator maps (Figure 3.1h), and inset maps, may

Figure 3.1   Map key elements.

also be included, depending on the context in which a given map is located.

3.2.1 Title, Scale, or Distance and Direction

It is important that a map bears a title because the title will invariably describe the major theme of the map. The scale or distance of the map is information that should also always be clearly indicated unless the intended audience is sufficiently familiar with the map's contents that the distance concept can be safely assumed. Maps have traditionally been compiled using a series of text scales, such as 1:20,000, 1:50,000, and so on. A scale bar may also be used, which may contain elements to assist the reader in locating a published map in a given series. Furthermore, a scale-bearing map assists in the cataloguing process and in helping individuals to identify those maps that might reveal the spatial patterns of, e.g., dengue fever cases, for example, because such maps exist at a scale that can show the necessary details.

The direction of the map is often related to the north, which indicates the map's orientation. On a given map, the north can be depicted symbolically in a similar way to the depiction of scale or distance. True north denotes the direction to the North Pole and is different from magnetic north, which changes because of variations in the earth's core and crust conditions. Often, the direction on a map indicates the true north; some maps might indicate both true and magnetic north.

3.2.2 Projection, Neat Lines, Locator, and Inset Maps

Projection is the process of transforming those places being mapped from the terrestrial sphere onto flat pages or a computer screen. In projection, two important properties can be identified: conformal property and equal area property. Conformal property ensures that small feature shapes are preserved on the projection. Here, the projection scale along the x and y planes is always equal; this property is important for navigation. Equal area property ensures that the measuring area on the map is in the same proportion to the corresponding area on Earth's surface, and this property is important for any analysis involving area. There are two systems of map projections: global and regional. Across Earth's surface, a global projection is used to define positions at all locations (e.g., GCS_WGS_1984); regional projection is used to define specific areas that often cover a country, state, or province (e.g., WGS_1984_UTM_Zone_36N).*

Neat lines or clipping lines are used to frame and indicate exactly where the area of a map begins and ends. They can also be used to clip the body of the map and of locator or inset maps. Some maps may portray areas that are unfamiliar to readers; therefore, a locator map can be useful for showing where the mapped data is in relation to an area that is familiar to the readers. Detailed maps are called *insets*. Sometimes, the data of mapped diseases is densely clustered within the small sections of a larger map, in relation to which the cartographer may wish to provide the reader with a close-up view of a location.

3.2.3 Legends and Symbology

Map legends define, interpret, and list the symbols used on a map. Symbols must appear in legends exactly as they appear in the body of the map. However, some maps do not require legends, especially simple maps for which the required information can be included in the caption.

In mapping, the data of vector-borne disease cases or transmitters can be classified as nominal, ordinal, and numeric. If attributes are given titles or names to distinguish one entity from another (e.g., Hassan as a patient name), the attributes are nominal, whereas if their values take on a natural order, the attributes are ordinal. For example, the location of dengue

---

* GCS means Geographical Coordinate System that enables every location on the Earth to be specified by a set of numbers or letters. WGS means World Geodetic System that comprises a standard coordinate system for the Earth, spheroidal reference surface for raw altitude data, and a gravitational equipotential surface that defines the nominal sea level. UTM means Universal Transverse Mercator. It is conformal projection that uses a 2-dimensional Cartesian coordinate system to give locations on the surface of the Earth.

cases may be classed in terms of risk, with class 1 representing high-, class 2 representing medium-, and class 3 representing low-risk locations. Attributes with numbers are numeric; examples might include the number of recorded disease cases in one location or temperature values. These values may be on a continuous scale or discrete, such as integers.

Points, lines, polygons, and other combinations of shape, hue, orientation, size, and texture may be used by cartographers to communicate the features of attributed data. In optimizing map interpretability, it is important to consider the selection of graphics to depict attributed spatial data and to decide how best to position such data on a given map. Nominal data can be represented by graphics and icons; this type of representation is apparently a simple matter, although in practice, automating placement to maximize clarity presents a range of analytical problems.

To position labels and symbols around geographical objects, most types of GISs have generic algorithms. Creating a window around text or symbols can help in positioning point labels to avoid overlap; splines can be used to label linear features and central points to assign area labels. Points and linear and polygon objects can be presented in the same manner with the property of the accommodated feature through the use of a hierarchy of graphic variables (e.g., lettering size and colour), especially when the attributes are ordinal.

When visualizing interval and ratio-scale attributed data, a number of conventions can be used (e.g., proportional circles and bar charts). These conventions are often used to assign interval or ratio-scale data to point or area locations. Variable line widths with increments that correspond to the precision of the representation continues the variation in flow diagrams. Variations in attributed data can also be represented by colour. One important issue in colour mapping is the selection of colours used in area shading. Colours have three dimensions: hue, saturation, and value. We commonly associate hue with colour tones such as red, blue, and purple, and these principally refer to distinguishing categories. Usually, a continuum of light-to-dark colour variation of a constant hue is used to represent low-to-high incidences of disease cases. *Value* refers to lightness or darkness; *saturation* denotes the predominant hue in the colour as well as its vividness.

In some health maps, the main aim is to show data values in relation to some average. For example, dengue fever cases rate above or below the national norm or z-score values. A diverging colour scheme is often highly recommended for this type of map. The choice of colours also depends on how, where, and by whom the map will be viewed; for example, many colours cannot be seen when viewed via an LCD (liquid crystal display) projector, others are difficult to reproduce in print, and so on. Although colour-blind people are able to see many hues, certain predictable groupings of hues can be confused with each other by such individuals, although this confusion depends on the severity of a person's colour vision deficiency.

3.2.4 *Dealing with Statistical Generalization*

Maps of vector-borne diseases are used to convey statistical information. A common procedure is to show the spatial patterns or trends (i.e., distribution) of disease (e.g., cases or vectors). This is usually summarized by some level of administrative area. Maps prepared for this task should depict the underlying distribution as unambiguously and accurately as possible. A balance needs to be struck between remaining true to the underlying data distribution and generalizing the data effectively to reveal the spatial pattern, especially when mapping vector-borne disease data.

As a rule, it is difficult to differentiate between more than five categories. Some basic classifications have been developed to divide continuous attributes. Changes to the class interval scheme can fundamentally change how maps look and the message they send. There is a range of options available to

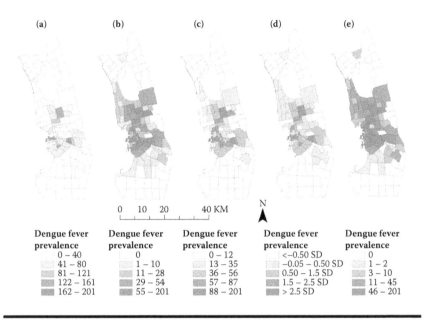

(a)   (b)   (c)   (d)   (e)

0   10   20   40 KM

N

Dengue fever
prevalence
  0 – 40
  41 – 80
  81 – 121
  122 – 161
  162 – 201

Dengue fever
prevalence
  0
  1 – 10
  11 – 28
  29 – 54
  55 – 201

Dengue fever
prevalence
  0 – 12
  13 – 35
  36 – 56
  57 – 87
  88 – 201

Dengue fever
prevalence
  <–0.50 SD
  –0.05 – 0.50 SD
  0.50 – 1.5 SD
  1.5 – 2.5 SD
  > 2.5 SD

Dengue fever
prevalence
  0
  1 – 2
  3 – 10
  11 – 45
  46 – 201

Figure 3.2   Choropleth maps showing dengue fever prevalence for 2010 using different classification methods: (a) equal, (b) quantile, (c) Jenks, (d) standard deviation, and (e) geometric.

cartographers or mapmakers to define class intervals. The first option is equal interval classification or break (Figure 3.2a). In this classification, the maximum and minimum data value ranges are divided into a fixed number of classes. This is useful for mapping malaria, Rift Valley fever, and dengue fever cases, for example, which follow a uniform distribution. It is not useful if there are extreme values in highly skewed data distributions, which can show the vast majority of areas falling into one or a few classes while others remain empty. Maps of such classifications will show little variation because most of the observed values are similar.

Second, quantile classification, also known as the n-tile method, can show an equal frequency of values in each of n classes (Figure 3.2b). Values are ranked from low to high, and these values are divided into a number n of classes. Each class contains an equal number of values. For instance, if n is 5 and a data set has 150 values, this will be divided into 5 classes with 30 values for each class. This classification option ensures each class is equally represented on the map. However, such results can be misleading because of the assignment of areas with similar data values to different classes. On the map itself, these areas will appear quite different even though their actual data values are close. This method is useful for highly skewed data distribution.

The Jenks method of natural breaks shows classes according to apparently natural groupings of data values (Figure 3.2c). The break definition is related to break points known to be relevant to a particular application (e.g., fractions and multiples of mean income levels, as well as rainfall thresholds known to support different levels of vegetation). The advantage of this method compared with others previously described is that it does not arbitrarily divide observations with similar values into different classes and produces class intervals that are either of equal width or with equal numbers of observation. Consequently, such results are unpredictable and dependent on the data.

Standard deviation and z-score classifications are often used when choropleth maps are compared (Figure 3.2d). In these cases, the distance of observation from the mean is shown. Each data value is transformed into a z score for a particular variable. Calculating the mean value is important in this context to generate class breaks in standard deviation measures above and below the mean. A minus 2.2 z score indicates that the data value is 2.2 standard deviations below the mean. In classifying maps using this method, odd numbers of class intervals with the middle interval number centred on zero can typically be created. This method is better used with attributed data that follows a normal distribution and lends itself to visual comparison because the map displays each variable in a common metric.

Arithmetic and geometric progressions are based on the width of the category intervals (Figure 3.2e). The category width intervals are increased in size at an additive (arithmetic) and multiplicative (geometric) rate. In arithmetic progressions, when the first category is 1-unit wide and the width is incrementally progressed by 1 unit, the second category will be 2 units wide, and so on (e.g., 1, 2, etc.). In the geometric progressions, when the first category is 3 units wide, the second category will be 9 units wide (3*3 = 9), and the third category will be 27 units wide (3*3*3 = 27). Both methods are useful for skewed distributions.

3.3 Mapping Vector-Borne Diseases

The importance of mapping vector-borne diseases at one or another spatial scale cannot be overstated, especially in the passage from 'not knowing' to 'knowing' or supposition to testable assumption and confirmed diagnosis. Vector-borne disease identification and treatment are essentially concerned with observing and then thinking about what we observe. Making maps of vector-borne diseases concerns a public reality wherein individual pathologies are transformed into public health events affecting communities and nations. Vector-borne diseases such as dengue fever or malaria epidemics and pandemics are spatial phenomena, and mapping them is how the disease threat is perceived and studied. Maps have become mechanisms by which the roles of the dead and the dying become shared realities whose relation to local environmental conditions can be assessed for decades. Mapping dengue fever or other vector-borne diseases that share a congress of symptoms also invites comparisons between patterns of disease incidence and local characteristics, such as congested housing and marshy swamps, that might either promote health or encourage a specific illness. For example, in the eighteenth century, mapping yellow fever in the US port cities showed the proximity of mortal cases to locations of noxious waste, which was subsequently used to develop a miasmatical theory of the disease.

It is important to consider vector-borne diseases on maps at every scale to understand these diseases and their histories. It is also important to study disease cases and their shared set of symptoms as evidenced across city and national maps. In mapping, boundaries are set and the content within those boundaries is organized. Such boundaries are essential to creating the context in which vector-borne disease theories are proposed and tested. Based on these perspectives, dengue fever and other vector-borne diseases are not only facts of experience, but also conclusions reached on the basis of investigations carried out within theoretical frameworks.

The GISs continue to transform the way that we investigate, map, and study vector-borne diseases. These are not paper based in the manner of traditional statistical, graphic, and cartographic approaches. The vector-borne diseases of our age occur at times when a GIS can provide useful tools for monitoring, controlling, predicting, mapping, and modelling these diseases because of its capability to capture, store, retrieve, manage, integrate, analyse, and display geographical and non-geographical data. GIS-based vector-borne disease occurrence or risk maps and models are increasingly used in endemic areas around the world. Using a GIS to map vector-borne diseases enlarges our thinking about the associations between microscopic agents, human host populations, and the climatic, environmental, and socioeconomic factors that encourage or inhibit their mutual association. In this scenario, a mapping-based GIS is not independent of but associated with the other tools of epidemiology and public health that help to increase our understanding of both the nature of human diseases and the potential for their containment.

The history of mapping medical issues generally and vector-borne diseases specifically is also the history of an ecological perspective on the emergence and control of conditions of these diseases insofar as mapping refers to a way of thinking that is inherently ecological. This approach allocates associations between factors of one or more abstract sets in a manner that permits all to be considered together. In these allocations, mapping can have at least one of the sets that thus joined has a spatial component that provides spatial data. This data is geographically descriptive. Furthermore, map thinking encourages a perspective that is at once rational and spatial. Creating maps shows the process by which map thinking is transformed into a concrete two-dimensional artefact (Eisen and Lozano-Fuentes, 2009). In this way, mapping is like telling a story, and creating a map is like writing and publishing it. In effect, creating maps transforms the story of the disease distribution into a form that can be produced and then shared.

3.3.1 Mapping and Modelling Malaria

One of the most dangerous vector-borne diseases worldwide is malaria, with cases occurring in more than 92 countries. *Plasmodium vivax, P. falciprum, P. malaria,* and *P. ovale* are the causes of human malaria. In tropical and subtropical countries, about 70 species are vectors of malaria under natural conditions; the major vectors are female *Anopheles* mosquitoes (Rogers et al., 2002; Capinha et al., 2009). According to the World Health Organisation (WHO, 2004), malaria has emerged as the most infectious killer of the tropical and subtropical diseases. Furthermore, many studies have used a GIS to describe malaria epidemiology.

Epidemiologists have used habitat suitability maps to study the distribution and abundance of several malaria species. These maps were mainly created in the fields of biogeography and conservation biology. The quantification of relations between the species and several environmental factors was the basis of these maps (Guisan and Thuiller, 2005). In mainland Portugal, Capinha et al. (2009) used values of former malaria distributions and suitable habitats to produce binary maps of suitable and unsuitable habitats for *Anopheles atroparvus.* These maps showed that former malaria distribution patterns were similar to *Anopheles* distribution in Portugal, suggesting that habitat maps of vectors can be good surrogates in the spatial assessment of malaria risk.

To formulate a timely and focused malaria control strategy, an information management system based on a GIS and using district and block-wise malaria data was constructed to map and highlight hot spots of malaria in Madhya Pradesh in India (Srivastava et al., 2009). The map displayed 58 blocks showing 25 districts as hot spots, helping the decision makers to keep those areas under intensive treatment. In addition, Bogh et al. (2007) georeferenced a Landsat Thematic Mapper (TM) image that contained 10 classes of land cover to UTM (Universal Transverse Mercator) zone 28, based on a 1:50,000 national survey map of Gambia, in which they digitized additional features, including national borders, main roads, villages, and the edges of floodplains. They then overlaid the mosquito data with the 10 classes of land cover and additional features to produce maps that showed the habitats representing suitable breeding areas for mosquitoes.

A GIS approach to map malaria risk in Africa was established by the South African Medical Research Council (SAMRC) with support from the International Development Research Centre (IDRC) (MARA, 1998). Statistical modelling, climatological charts, and fieldwork were included in this project to produce a state map of malaria. In addition, the progress and current status of the GIS with reference to *P. falciparum* malaria in sub-Saharan Africa was reviewed (Hay and Lennon, 1999). Hay et al. (2000) focused on the ecology of *P. falciparum* and its major *Anopheles* vectors to provide a background for the study of transmission processes and their environmental correlates. In addition, because of the lack of spatially defined data and a clear understanding of how epidemiological variables relate to disease outcomes, the limited use of epidemiological maps in malaria control in Kenya was noted (Omumbo et al., 1998).

In India, villages with conditions conducive to the incidence of malaria were determined using topographical maps, satellite-generated maps, and ArcInfo software (Shirayama et al., 2009). A composite map featured 13 stratification classes by sequentially integrating environmental factors such as hydro-geomorphology, water table level, water quality, soil type, and relief and irrigation channels. This study found that malaria incidence is mainly related to water table, soil type, irrigation, and water quality (Srivastava et al., 2009). In China, particularly in the Yunan Province, a GIS and multiple regressions were used to determine the nature and extent of factors influencing malaria transmission. Data from 1990–1996 was collected and analysed; the results showed that the combined physical environmental effects, the presence of compatible vectors, and the degree of population mobility influenced the malaria situation (Hu et al., 1998).

3.3.2 Mapping and Modelling Dengue Fever

Dengue fever is caused by a family of viruses (DEN-1, –2, –3, –4) that is transmitted by mosquitoes. It is an acute disease of abrupt onset that usually follows a benign course with headache, fever, exhaustion, severe joint and muscle pain, swollen glands, and rash (Achu, 2008).

DF attacks people who have low levels of immunity. Hence, it is possible to contract DF multiple times. An attack of DF produces immunity for a lifetime only to that particular type to which the patient was exposed (Cunha, 2007; WHO, 2002). Currently, DF is prevalent throughout tropical and subtropical regions around the world, predominantly in urban and semi-urban areas. Outbreaks have occurred in South and Southeast Asia, the Caribbean, the US Virgin Islands, Cuba, Central America, Australia, and Saudi Arabia.

A GIS has been applied in DF research in a number of studies. For example, Barrera et al. (2000) investigated the stratification of a city with hyperendemic dengue haemorrhagic fever (DHF) transmission to identify hot spots for the application of surveillance and control measures. A GIS has

also been applied to analyse economic resources and diseases with reference to DF and malaria in Thailand. Provincial
products and health care resources in relation to geographical
distribution were examined in this study. The authors argued
that overall planning can be carried out at national and mul-
ticountry levels; they also acknowledged that the disease data
and socioeconomic data was collected at different times and
in different ways, which thus limited dynamic interpolation of
the two data sets (Indaratna et al., 1998).

Tran et al. (2004) used the Knox test, a classic space-time
analysis technique, to detect spatiotemporal clustering and
demonstrated the relevance and potential for the use of a
GIS and spatial statistics for the elaboration of a dengue fever
surveillance strategy. Strickman and Kittayapong (2002), using
spatial analysis, identified locations with higher concentrations of the dengue vectors. To visualise and map the effect
of open marsh water management for mosquito vector control
when merged with invasive plant and salt marsh restoration, a
GIS was also used. The results showed a significant reduction
in the frequency of finding larvae on the marsh surface, leading to loss of spatial larval hot spots in the area under open
marsh water management (Rochlin et al., 2009). Morrison et al.
(1998) carried out a space-time analysis of reported dengue
cases during an outbreak in Florida between 1991 and 1992.
Pratt (2003) found that incorporating traditional epidemiological statistical techniques into a GIS interface allows researchers to gain greater insight into the spatial aspect of the spread
of disease.

In their study of the use of a GIS in ovitraps monitoring for
dengue control in Singapore, Tan and Song (2000) developed
three models to monitor, analyse, and evaluate ovitrap breeding data. Their goal was to better understand the *Aedes* situation on the island for planning vector surveillance and control
operations. Household surveys of dengue infection during
2001–2002, spatial point pattern analysis, and risk factor
assessments were used to illustrate the spatial heterogeneity

in the risk areas of dengue when using a spatial approach in a short time interval. The results indicated that the low-prevalence areas in 2001 had shifted to high-risk areas the following year (Siqueira et al., 2008). Schafer and Lundstrom (2009) used the geographical distribution of *Aedes sticticus* and climate change data to model the future distribution of this vector. The model showed that the *Aedes sticticus* potential areas with suitable conditions would likely increase. An information value method in the GIS environment was used to analyse and obtain the influence of physioenvironmental factors, such as land use and land cover, on the incidence of DF (Nakhapakorn and Tripathi, 2005).

Chansang and Kittayapong (2007) integrated immature sampling methodology with GIS technology to produce spatial density distribution maps and to identify the clusters of immature stages and breeding sources for improving the surveillance and control systems of *Aedes aegypti*. This study found that water jars of various types and cement bath basins were the two main breeding sources. In the state of Hawaii, geographic analysis and GIS spatial/temporal analysis were conducted on the 2001–2002 DF outbreaks to create the dengue threat model (Napier, 2001). Takumi et al. (2009) assessed whether *Aedes albopictus* that were found at Lucky Bamboo import companies in the Netherlands could produce subsequent generations. Based on a GIS and collected climatic variables data, they located suitable and unsuitable regions for the *Aedes albopictus* species.

In Rio de Janeiro, Brazil, maps of *Aedes aegypti* density were generated using the Infestation Index obtained from the *Aedes aegypti* Infestation Index Rapid Survey, the Breteau Index, and spatial pattern analysis. The map represented five areas with high and medium density of positive *Aedes aegypti* breeding locations and highlighted small block clusters with high larvae density. Chaikoolvatana et al. (2007) aimed to develop a GIS for *Aedes aegypti* surveillance and dengue haemorrhagic fever in northeastern Thailand. Their development proceeded via three stages: (1) collecting primary and secondary data such as those related to dengue vector incidence, water storage containers, and number of reported dengue haemorrhagic fever cases/100,000 population; (2) analysing the data; and (3) searching the target location and presenting the results via figures on maps. The increase in the number of dengue haemorrhagic fever cases during high disease incidence suggested a strong correlation between the peak rainfall, the high density of *Aedes aegypti* mosquitoes, and the high incidence of DHF cases. In Singapore, it was assumed that the *Aedes aegypti* mosquito breeds indoors and people are infected in their homes; however, a GIS showed that the groups most infected by DF and DHF were mobile, such as teenagers and young adults who spent most of the time outdoors. The surveillance systems were changed to fortnightly checks on outdoor areas, the result of which showed a decline in the number of dengue fever and dengue haemorrhagic fever cases (Tan, 2001). This study is an example of how a GIS can lead to an outcome that diminishes the prevalence of a given disease.

3.4 Conclusion

Maps of vector-borne diseases play an important role in descriptive spatial epidemiology. Understanding cartography and the creation of maps is useful in identifying areas under disease risk, helping to formulate hypotheses about disease aetiology, and assessing health resource allocations. The information displayed on maps can also be used for providing treatment, monitoring vectors and cases, and preventing incidence of disease. Furthermore, maps can be used to verify disease hot spots as data is being collected and to target those hot spot locations suitable for spraying and eliminating disease vectors, which is another key preventive measure. Finally, both cartographer and mapper should carefully consider the selection of study area boundaries, scale, geographic projection, and symbology. The linking of maps with other forms of data display and dealing with statistical generalisations are aspects of visualization that can contribute to the achievement of these objectives.

References

Achu, D.F. (2008). Application of GIS in Temporal and Spatial Analyses of Dengue Fever Outbreak: Case of Rio de Janeiro, Brazil. Master's thesis, Linköpings University.

Barrera, R., Delgado, N., Jimenez, M., Villalobos, I., Romero, I. (2000). Stratification of a hyperendemic city in hemorrhagic dengue. *Revista Panamericana de Salud Publica*, 8(4): 225–233.

Bogh, C., Lindsay, S.W., Clarke, S.E., Dean, A., Jawara, M., Pinder, M., Thomas, C.J. (2007). High spatial resolution mapping of malaria transmission risk in the Gambia, West Africa, using Landsat TM satellite imagery. *American Journal of Tropical Medicine and Hygiene*, 76(5): 875–881.

Capinha, C., Gomes, E., Reis, E., Rocha, J., Sousa, C. A., do Rosario, V.E., Almeida, A.P. (2009). Present habitat suitability for *Anopheles atroparvus* (Diptera, Culicidae) and its coincidence with former malaria areas in mainland Portugal. *Geospatial Health*, 3(2): 177–187.

Chaikoolvatana, A., Singhasivanon, P., Haddawy, P. (2007). Utilization of a geographical information system for surveillance of *Aedes aegypti* and dengue haemorrhagic fever in north-eastern Thailand. *Dengue Bulletin*, 31: 75–82.

Chansang, C., Kittayapong, P. (2007). Application of mosquito sampling count and geospatial methods to improve dengue vector surveillance. *American Journal of Tropical Medicine and Hygiene*, 77: 897–902.

Eisen, L., Lozano-Fuentes, S. (2009). Use of mapping and spatial and space-time modeling approaches in operational control of *Aedes aegypti* and dengue. *PLoS Neglected Tropical Diseases*, 3(4). doi:10.1371/journal.pntd.0000411

Guisan, A., Thuiller, W. (2005). Predicting species distribution: offering more than simple habitat models. *Ecology Letters*, 10: 19.

Hakre, S., Masuoka, P., Vanzie, E., Roberts, D.R. (2004). Spatial correlations of mapped malaria rates with environmental factors in Belize, Central America. *International Journal of Health Geographics*, 3(6). doi:10.1186/1476-072X-3-6

Hay, S.I., Lennon, J.J. (1999). Deriving meteorological variables across Africa for the study and control of vector-borne disease: a comparison of remote sensing and spatial interpolation of climate. *Tropical Medicine and International Health*, 4(1): 58–71.

Hay, S.I., Omumbo, J.A., Craig, M.H., Snow, R.W. (2000). Earth observation, geographic information systems and *Plasmodium falciparum* malaria in sub-Saharan Africa. *Advances in Parasitology*, 47: 173–215.

Hu, H., Singhasivanon, P., Salazar, N.P., Thimasarn, K., Li, X., Wu, Y., … Looarecsuwan, S. (1998). Factors influencing malaria endemicity in Yunnan Province, PR China (analysis of spatial pattern by GIS). Geographical Information System. *Southeast Asian Journal of Tropical Medicine and Public Health*, 29(2): 191–200.

Indaratna, K., Hutubessy, R., Chupraphawan, S., Sukapurana, C., Tao, J., Chunsutthiwat, S., … Crissman, L. (1998). Application of geographical information systems to co-analysis of disease and economic resources: dengue and malaria in Thailand. *Southeast Asian Journal of Tropical Medicine and Public Health*, 29: 669–684.

Khormi, H., Kumar, L. (2011). Examples of using spatial information technologies for mapping and modelling mosquito-borne diseases based on environmental, climatic, socio-economic factors and different spatial statistics, temporal risk indices and spatial analysis: a review. *Journal of Food, Agriculture and Environment*, 9(2), 41–49.

Khormi, H.M., Kumar, L. (2012). Assessing dengue fever risk based on socioeconomic and environmental variables in a GIS environment. *Geospatial Health*, 6(2), 171–176.

MARA. (1998). *Towards an Atlas of Malaria Risk in Africa. First Technical Report of the MARA Collaboration*. Durban, South Africa: MARA.

Morrison, A.C., Getis, A., Santiago, M., Rigau-Perez, J.G., Reiter, P. (1998). Exploratory space-time analysis of reported dengue cases during an outbreak in Florida, Puerto Rico, 1991–1992. *American Journal of Tropical Medicine and Hygiene*, 58(3), 287–298.

Nakhapakorn, K., Tripathi, N.K. (2005). An information value based analysis of physical and climatic factors affecting dengue fever and dengue haemorrhagic fever incidence. *International Journal of Health Geographic,* 4: 13–15.

Napier, M. (2001). Application of GIS and modeling of dengue risk areas in the Hawaiian islands. Retrieved from http://www.pdc.org/PDCNewsWebArticles/2003ISRSE/ISRSE_Napier_TS49.3.pdf

Omumbo, J., Ouma, J., Rapuoda, B., Craig, M.H., le Sueur, D., Snow, R.W. (1998). Mapping malaria transmission intensity using geographical information systems (GIS): an example from Kenya. *Annals of Tropical Medicine and Parasitology,* 92(1), 7–21.

Pratt, M. (2003). Down-to-earth approach jumpstarts GIS for dengue outbreak. *Magazine for Esri Software Users,* 6: 2.

Robinson, A. (1982). *Early Thematic Mapping in the History of Cartography.* Chicago: University of Chicago Press.

Rochlin, I., Iwanejko, T., Dempsey, M. E., Ninivaggi, D. V. (2009). Geostatistical evaluation of integrated marsh management impact on mosquito vectors using before-after-control-impact (BACI) design. *International Journal of Health Geographics*, 8: 1–20. doi:10.1186/1476-072x-8-35

Rogers, D.J., Randolph, S.E., Snow, R.W., Hay, S.I. (2002). Satellite imagery in the study and forecast of malaria. *Nature*, 415(6872): 710–715.

Schafer, M.L., Lundstrom, J.O. (2009). The present distribution and predicted geographic expansion of the floodwater mosquito *Aedes sticticus* in Sweden. *Journal of Vector Ecology*, 34(1): 141–147. doi:10.1111/j.1948-7134.2009.00017.x

Shirayama, Y., Phompida, S., Shibuya, K. (2009). Geographic information system (GIS) maps and malaria control monitoring: intervention coverage and health outcome in distal villages of Khammouane province, Laos. *Malaria Journal*, 8: 217.

Siqueira, J.B., Maciel, I.J., Barcellos, C., Souza, W.V., Carvalho, M.S., Nascimento, N.E., … Martelli, C.M.T. (2008). Spatial point analysis based on dengue surveys at household level in central Brazil. *BMC Public Health*, 8. doi:10.1186/1471-2458-8-361

Snow, R., Gouws, E., Omumbo, J., Rapuoda, B., Craig, M., Tanser, F., Sueur, D., Ouma, J. (1998). Models to predict the intensity of *Plasmodium falciparum* transmission: applications to the burden of disease in Kenya. *Transactions,* 92: 601–606.

Srivastava, A., Nagpal, B.N., Joshi, P.L., Paliwal, J.C., Dash, A.P. 2009. Identification of malaria hot spots for focused intervention in tribal state of India: a GIS based approach. *International Journal of Health Geographics*, 8: 30.

Strickman, D., Kittayapong, P. (2002). Dengue and its vectors in Thailand: introduction to the study and seasonal distribution of *Aedes* larvae. *American Journal of Tropical Medicine and Hygiene*, 67: 247–259.

Tan, A., Song, R. (2000). The use of GIS in ovitrap monitoring for dengue control in Singapore. *Dengue Bulletin*, 24: 110–116.

Tan, B.T. (2001). New initiatives in dengue control in Singapore. *Dengue Bulletin*, 25: 1–6.

Takumi, K., Scholte, E.J., Braks, M., Reusken, C., Avenell, D., Medlock, J.M. (2009). Introduction, scenarios for establishment and seasonal activity of *Aedes albopictus* in the Netherlands. *Vector-Borne and Zoonotic Diseases*, 9(2): 191–196. doi:10.1089/vbz.2008.0038

Tran, A., Deparis, X., Dussart, P., Morvan, J., Rabarison, P., Remy, F., … Gardon, J. (2004). Dengue spatial and temporal patterns, French Guiana, 2001. *Emerging Infectious Diseases*, 10(4): 615–621.

World Health Organisation (WHO). (2004). Roll Back Malaria. Retrieved from http://who.int/int-fs/en

World Health Organisation (WHO). (2002). Dengue and dengue haemorrhagic fever. Retrieved from http://www.who.int/media-centre/factsheets/fs117/en/index.html

Chapter 4

Spatial Data

4.1 Introduction

Spatial data is information that links the feature to a geographic location; that is, it links the feature to a position on Earth. Therefore, spatial data contains geographic information that enables specifying exactly where on Earth that feature is located. Because of this, spatial data is often also termed geospatial or georeferenced data. This ability to link the feature to an exact location on Earth is extremely important as it enables building relationships between that feature and other features around it. For example, a person suffering from dengue fever visits a doctor and is asked for the address where they live. This address now enables the doctor and anyone else using this data to see exactly where on Earth this person lives and to investigate the surrounding environmental conditions. The address is the spatial data and is denoted by coordinates on a map (Figure 4.1).

Spatial data is stored as coordinates and topology and is most often accessed, manipulated, or analysed through geographic information systems (GIS).

The traditional method for storing, analysing, and presenting spatial data is the map. For example a two-dimensional (2D) road map contains points, lines, and polygons that represent

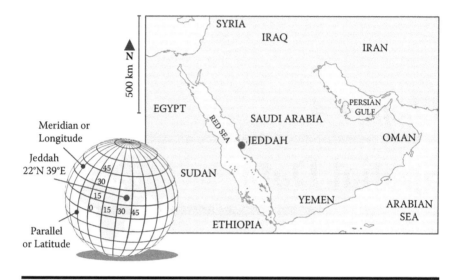

Figure 4.1 Representation of spatial data on a map.

different objects, such as cities, roads, and district boundaries. The features in the map have specific locations and areas and visualise geographic information; therefore, they are considered spatial data. So, understanding maps and the way they are produced is essential for exploring the characteristics of spatial data (Heywood et al., 2002).

Geographic data have two components: spatial and non-spatial. The spatial component contains the locational information; the nonspatial component contains attribute information, commonly called descriptive information. As an example, the dengue patient's name, date of birth, age, marital status, and so on all become part of the nonspatial or attribute information. Figure 4.2 shows an example of how spatial data are related to nonspatial data.

Spatial features may be discrete (also referred to as thematic, categorical, or discontinuous data) or continuous (also referred to as field, nondiscrete, or surface data). Discrete objects have known and definable boundaries, and they do not exist between observations and form separate entities (Dent, 1999). For example, a house on a farm is a discrete object within the surrounding landscape (farm). The house

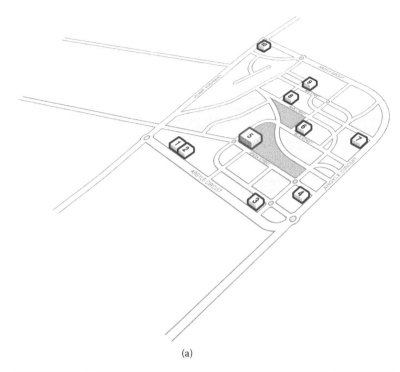

(a)

Feature ID	Shape	Patient Name	Address	Age	Year of Birth	Gender	Marital Status
1	Point	Gaffoor	Argyle Cct	47	1967	M	M
2	Point	Ahmed	Argyle Cct	22	1992	M	S
3	Point	Khan	Argyle Cct	35	1979	M	M
4	Point	Prasad	Argyle Cct	32	1982	M	M
5	Point	Ismail	Green St	8	2006	M	S
6	Point	Singh	Blue St	67	1947	M	M
7	Point	Ali	Argyle Cct	55	1959	M	M
8	Point	Zaid	Main Cct	39	1975	M	S
9	Point	Nisha	Red St	6	2008	F	S
10	Point	Albalawi	Hume Hwy	31	1983	F	M

(b)

Figure 4.2   Relationship between spatial and nonspatial (attribute) data: (a) points showing dengue fever patient locations; (b) related attribute information.

edges can be definitively established. However, a continuous surface or data exists spatially between observations (Dent, 1999) and represents phenomena in which each location on the surface is a measure of the concentration level or its relationship from a fixed point in space. Elevation, aspect, and precipitation are examples of continuous data.

4.2 The Importance of Spatial Data

Spatial data and geographically referenced information are crucial for subjects of modern research, including disease mapping and analysis, urban and rural developments, environmental protections, natural and technological disaster planning and management, and health risk assessments. Many social, economic, and political practical decisions on a daily basis rely on spatial information (Onsrud and Rushton, 1995). For example, accurate information (e.g., address, location of services, population, climate, and topography) is fundamental for directing disaster management and emergency services (McDougall et al., 2007). Sources of spatial data are generated with a purpose in mind, such as turning data into information that can be used in decision making by a third party. For example, spatial data sets on a modelled global disease and vector distributions provides relevant information on the risks of vector-borne disease in the air travel network that can be used by passengers, travel planners, airlines, and biosecurity agencies. Spatial data has proven crucial in detecting, mitigating, preparing for, responding to, and recovering from disease epidemics. The increased availability of spatial environmental, health, and population data combined with different statistical methods and spatial analysis techniques has provided a powerful tool in evaluating the spatial relationships between disease and environmental hazards (Beale et al., 2008).

Within the context of the rapid growth and the cost of data capture and management, the interoperability of spatial data enables more efficient management and saves significant time and money. Commonly, these data sets do not reside within one organization; therefore, they might need to be shared between different organizations. Sharing and reusing data not only can reduce costs but also can improve data quality by individuals who validate the data (Williamson et al., 2003). Although there is a history of good cooperation between local, state, and national organizations, sharing of data can be problematic (McDougall et al., 2007). Therefore, spatial data infrastructures (SDIs) have been developed to encompass the efficient and flexible collation, management, access to, and use of spatial data. With SDIs, information can be used across a wide range of processes and applications beyond the original intent of the data.

4.3 Characteristics of Spatial Data, Including Topology and Topological Relationships and Their Importance

Spatial data are described by their spatial characteristics (point, line, polygon, and surface) and attribute characteristics (position, length, area, extent). Spatial data are believed to be indefinite because of the different representation of each spatial object's content and structure (Altibase, 2014). The continuous geographic and map elements must be dispersed and abstracted in discrete data to identify and dispose of geographic elements (Cui, 2001). There are two conventional spatial data models: field based and entity based. The field-based model is employed in the representation of any space using a series of attributes, and the entity-based model represents space as an object (or entity) made up of a set of points (Altibase, 2014). Usually, the most important objects in the entity-based model are points, lines, and polygons. Each of these basic spatial entities is a simple 2D model that can be used to represent real-world features.

4.3.1 Points

Points are spots that have no physical and actual spatial dimensions but have specific locations and are used to represent features that are too small to be represented as areas

(Davis, 2001; Heywood et al., 2002). For example, locations of mosquito traps or infected persons' houses can be shown as points on a map. The points, however, do not indicate the actual size, length, or shape of the feature. The data stored for these points includes geographical location and details about the feature. Such point features are not fully described by a 2D geographical reference (x, y) as there is always a height component because these features are located at some height above sea level.

4.3.2 Lines

Lines, which indicate an ordered set of points, are used to represent linear features in the real world or features that do not exist in real life but are demarcated, such as administrative boundaries (Heywood et al., 2002). Lines are ordered strings of x, y coordinates joined together. Lines may be isolated or connected in networks, such as utility and drainage lines. Different line thicknesses (line weights) can be used to indicate different classes of a feature, such as road types.

4.3.3 Polygons

A polygon is a collection of lines that forms a closed loop, a 2D feature with at least three sides that has an area and perimeter (Davis, 2001). Area entities are often referred to as polygons. A polygon may be a simple single connected area or a complex collection formed by some embedded simple polygons, such as the states of a country.

4.3.4 Surface

A three-dimensional (3D) area is considered a surface. Surfaces are used to represent topography or nontopographical variables such as population densities.

Table 4.1 gives some examples of each feature type.

Table 4.1   Feature Types and Example Applications

Feature Class	*Application and Use*	*Examples*
Point	Point locations	Mosquito traps, address of a dengue patient
Line	Linear features	Roads, rivers, drains with mosquito larvae, sewer lines
Polygon	Area features	Lakes, district with high dengue infection rate, dense forest stand suitable for mosquito breeding
Surface	3D areas	Elevation data

4.4 Relationship between Entities

A key function of modern GIS packages is in the arena of spatial analysis. The analysis and organization of large quantities of spatial data and understanding the relationships between the features are parts of a critical role of a GIS. Spatial relationships show how objects relate to each other in space. Spatial relationships include the following (Davis, 2001; Sahu, 2007):

- Location of features, which is usually defined based on standard geographic coordinate systems, such as latitudes and longitudes
- Distance between features, which is determined through simple measurements and trigonometry
- Distribution or clustering, which is the collective location of features as range or geographical dispersal
- Proximity of features to each other or some other feature of interest
- Density, which is the number of items in a unit area; shows how close the features are from each other
- Pattern, which is the arrangement of features in space
- Direction of movement; also relates to temporal analysis
- Boolean relationships

4.5 Topology

Topology is the term used to describe the geometric characteristics of objects in which the spatial relationships between
features are expressed explicitly (Chang, 2002). Topology links
the attribute and spatial data and helps in understanding the
relationship of the feature with neighbouring features. This
element is critical because of the need to explore the spatial relationships between features in the landscape. It is also
useful for detecting digitizing errors in digital maps, as well
as overlaying operations and network analysis (Chang, 2002).
The topological characteristics of an object are independent of
scale of measurement (Chrisman, 1997). Topology, as related
to spatial data, consists of three elements: adjacency, containment, and connectivity (Burrough, 1986). Adjacency (sharing
common boundaries or nearness) and containment (areas
contained within another area or the degree of connectivity)
describe the geometric relationships between the areas of features (Heywood et al., 2002). Connectivity is used to describe
the linkage or connection between line features (shared node
in arc-node topology) (Sahu, 2007).

When topology is applied to GIS, data structure tables
are built for features. These tables determine various relationships, and by utilizing these tables, GIS software can
track connections and measure distances and areas (Davis,
2001). In GIS, topology offers special information, such as
length, distance, and area of the data, and provides powerful functions of spatial analysis. It also creates connections
between features, and each feature can be linked to several
other features (multiple linkages or connections). All the
functional connections, distances, and other spatial relationships can result in an ideal interpretation of network features
and performance of specialized analysis (network analysis)
(Davis, 2001).

4.6 Georeferencing

Before data may be used in GIS the data must have a real-world coordinate system so that they become suitable geographic data. In essence, map coordinates are used to assign a spatial location to the data or map features. This process is termed *georeferencing*. The assignment of spatial locations enables the data to be viewed, queried, and analysed with other geocoded geographic data (Figure 4.3). There are numerous georeferencing systems that describe the real world in different ways and with varying degrees of precision for different regional or national purposes (Bernhardsen, 2002). The only true geographic coordinates are latitude (measured in degrees north or south of the equatorial plane) and longitude (measured in degrees east or west of the Greenwich meridian). The location of any point on Earth's surface can be defined by a reference using latitude and longitude.

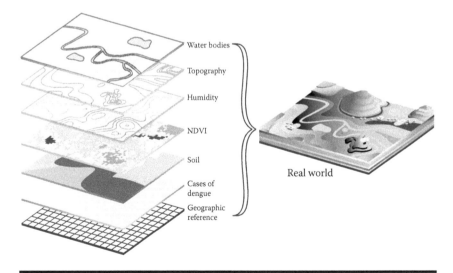

Water bodies

Topography

Humidity

NDVI

Soil

Cases of dengue

Geographic reference

Real world

Figure 4.3   Georeferenced map layers align perfectly so that information from multiple layers links.

4.7 Spatial Referencing

Data in a GIS are associated through the use of a spatial referencing system. A spatial reference system includes the map projection, datum, and coordinate system and enables one to interrogate data from different coordinate systems.

4.7.1 Map Projections

Even when a map data file has been georeferenced to a specific coordinate system, there is still a need for the GIS to know which map projection to use to give proper spatial characteristics to features (Davis, 2001; Johnson, 2011). The location of features on Earth's surface is represented as a map feature on a plane surface. The systematic transformation of the longitude and latitude of features from the spherical geographic grid map to a Cartesian coordinate system (plane) is known as map projection. In simple terms, it is the transformation of the spherical or ellipsoidal Earth onto a flat map. Map projection is important for GIS users because if different databases are presented with different map projections, it will result in inaccurate analyses and map features will not register with another spatially (Chang, 2002). On a globe, features from Earth's surface (their shape, the area they occupy, and the distance and direction between them) are correctly shown. An ideal map projection retains all of these characteristics and translates them to the map. So, an ideal map would have the following characteristics:

1. Conformality: The retention of the correct shape of Earth features on the map.
2. Equivalence (Equal Area): A unit area on a map represents the same number of square kilometres on Earth's surface.

3. Distance: The length of a straight line between two points on the map represents the correct circle distance between the same points on Earth.
4. Direction: A straight line drawn between two points on the map shows the correct azimuth of the line.

There are many projections, each with its own advantages and disadvantages. None of the map projections can maintain all of the important spatial variables, such as shape, direction, and distance (Johnson, 2011). As shown in Figure 4.4, the projections are classified into three groups—cylindrical, conical, and azimuthal—based on the underlying geometrical

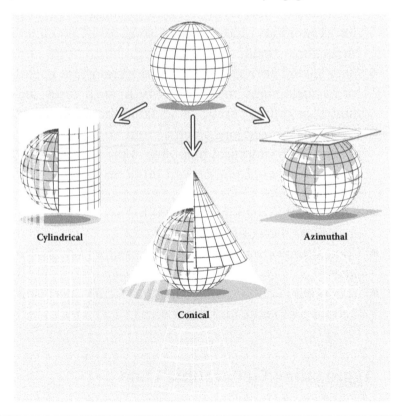

Figure 4.4   Azimuthal, conical, and cylindrical map projections.

conversions and can be unfolded to one plane and function as a planar map (Bernhardsen, 2002).

Cylindrical, conical, and azimuthal projections can be made equidistant, equal in area, or conformal. It is also possible to convert to and from different projections based on the mathematical functions used in GISs without loss of accuracy (Johnson, 2011).

4.7.2 Selecting a Map Projection

1. Purpose of the map
   - Maps that show the correct relationship between areas are useful for displaying political boundaries, distribution commodity, and so on. In such cases, equal area projections should be used.
   - Maps giving correct angles and directions are useful for navigators and meteorologists. In such cases, azimuthal or orthomorphic projections should be used.
   - Conformal projections, such as transverse mercator, are used for topographical mapping; short distance, direction, and area calculation can be done on these maps.
2. Area on the globe
   - Cylindrical projections are used for tropical areas or areas near the equator.
   - Conical projections are generally used for temperate areas.
   - Azimuthal/zenithal projections are usually more suitable for temperate and polar areas.

4.8 Aggregating Geographic Data

The grouping of spatial data at a level of detail or resolution coarser than the level at which the data was collected (at finer resolutions) is called *aggregation*. So, during an aggregation process, the original spatial data is reduced to a smaller

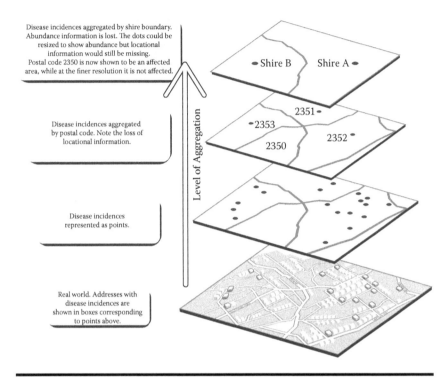

Disease incidences aggregated by shire boundary. Abundance information is lost. The dots could be resized to show abundance but locational information would still be missing. Postal code 2350 is now shown to be an affected area, while at the finer resolution it is not affected.

Disease incidences aggregated by postal code. Note the loss of locational information.

Disease incidences represented as points.

Real world. Addresses with disease incidences are shown in boxes corresponding to points above.

Level of Aggregation

Shire B   Shire A

2351
2353
2350   2352

Figure 4.5   Loss of fine-scale locational information because of data aggregation.

number of data units; hence, each aggregated data unit represents an area larger than the original units (Figure 4.5). Data aggregation is widely reported as "scaling up" in different environmental studies and models from local to regional or global scales (Blan and Butler, 1999). The outcome from aggregation is a loss of spatial and attribute detail (i.e., outputs differ when input data of different resolutions is used) (Rao, 2012). There are various reasons for aggregation of spatial data, including generating data, generalizing/summarizing data, simplifying maps, updating databases, so analysis is required at a different scale and needs to match with data from different sources that are at different spatial resolutions. A GIS facilitates aggregation through a variety of techniques (e.g., dissolve, merge, spatial join, resample) and evaluates issues related to the use of aggregate data (Rao, 2012). In GIS "dissolve" aggregates features based on specified attributes (merging adjacent polygons, lines, or regions that have the same value for an attribute item). For example, in dissolve, an area where dengue is prevalent could be merged with another area where malaria is prevalent, creating a new and larger area that could be labelled as a disease-prone area. Aggregation is accomplished through selection of a distance criterion or containment (Kemp, 2008). The user can decide which attributes to be summarized (summary statistics) during aggregation (e.g., averages, sums, weighted averages).

4.9 Data Types

GIS and environmental models function with a broad spectrum of geospatial data used for spatial analysis. These data are in different formats, from different sources, and are captured using different measurement scales. Data may be categorized into several different types, including spatial data, attribute data, temporal data, and metadata. Spatial data is divided into two types, vector and raster. Attribute data provide critical information about the spatial objects. Temporal data represent aspects of time. For example, temperature and wind speed, which change over time, are considered temporal data. Metadata is information about data, which is useful when data are exchanged between users. Metadata generally includes information about projections, source material, and how data were processed.

4.9.1 Raster and Vector Data Models

Drawings, paper maps, or a GIS are all models of reality. In each of these, we try to show what the real world (or some part of it) looks like. However, the real world is too complex to show every detail. Take, for example, a paper map showing the drainage network of Australia. It definitely will have all the major rivers, maybe the intermediate and minor rivers, but it may not be possible to show all small creeks on this map. If an attempt is made to show all the small creeks and drains, there is too much information on the map to gain any meaningful insight.

Also, if you tried to show every detail of the world in a GIS, you would run out of time collecting the information and the computer would run out of storage space. Therefore, you only collect those parts of reality that you believe are important for your research application and only in the detail necessary. This obviously leads to some problems, which we discuss further elsewhere.

The data that you collect needs to be represented in some form in the computer. A GIS uses two basic data formats to represent spatial features: vector and raster. A vector data model uses points and their x, y coordinates (real-world entities) to construct spatial features of points, lines, and areas, whereas the raster data model uses a grid cell structure. These data models determine how the data is structured, analysed, stored, and processed in the GIS.

4.9.2 Raster Data Model

The raster model divides space into an array of regularly spaced square cells, sometimes called pixels. It displays, locates, and stores graphical data using a matrix of grid cells or pixels. The entire study area is divided into a regular grid of cells, with each cell containing a single value that represents the feature within the cell area. In the raster model, a unique reference coordinate represents each pixel. Each cell in the grid has two associated values: a positional value that marks its identity and an attribute value of the underlying area it represents (e.g., elevation, land use, etc.). It is important to note that the raster model deals with locations, not objects per se. Note that you need not have only one feature per raster map layer. You can have a number of features, but each feature will have a different code (value). Each cell in the grid contains a value or identity corresponding to the information and characteristics of the spatial feature at that location. The cell is the minimum mapping unit as it is the smallest size by which a landscape can be presented. The cell does not maintain the true size, shape, and location of the features. Digital aerial photographs, satellite images, digital pictures, and scanned maps are all raster layers. Because the raster model stores data for each location of the area of interest irrespective of whether data of interest is available for that point, it is ideal for describing spatially continuous data, such as elevation, biomass, temperature, and rainfall. Figure 4.6 shows the structure of a raster grid; Figure 4.7 provides an example of raster data representation.

It should be obvious from this example that one grid cell is one unit and holds one feature only. In addition, every cell has a value, even if it is 'missing'. In the example, this is denoted by the 0s. A cell can hold a number or an index value standing for an attribute, and each cell has a resolution. The resolution is given as the cell size in ground units.

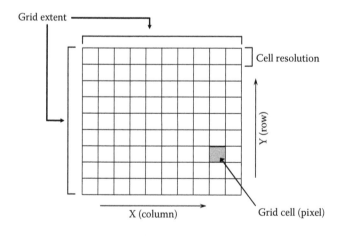

Figure 4.6 Structure of a raster grid.

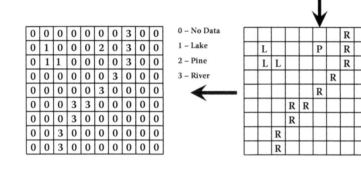

Figure 4.7   Raster data representation.

4.9.3 *Vector Data Model*

In GIS, the vector is the map-like drawing of features without the pixels or grid cells. The vector model is based on directed line segments, called *vectors*. The vectors are used to build a complex representation (Figure 4.8). The features in a vector model are based on points, which are used to show discrete locations by using pairs of spatial coordinates: lines, which are a set of coordinates interconnected to represent a linear shape, and polygons, which are homogeneous areas formed by a number of line segments forming a closed loop. Shapes are better retained and spatially accurate in vectors compared to rasters because the lines are not broken into cells and continue from the start to the end in a continuous manner (Davis, 2001). In GIS, vector features are defined by their shapes or the outline of their shapes. The vector model is also known as

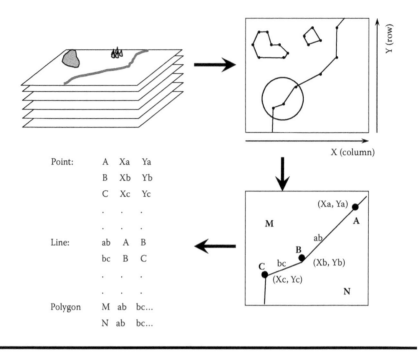

Figure 4.8   Vector data representation.

an object-oriented or object-based system because information is stored by objects rather than discrete locations.

The vector system is useful for describing distinct features in the landscape (roads, property boundaries, etc.). So, points would represent locational features such as wells, bore holes, telephone poles, mosquito traps, infected person's house location, and so on; lines would represent linear features such as roads, streams, pipelines, drains with mosquito larvae, and so on; and polygons would represent area features such as lakes, census boundaries, land ownership, areas with high rainfall prone to mosquitoes, and so on.

4.9.4 So, Which Data Model to Select?

Although these two models are used in GIS, it can be difficult for the user to know which to request and work with. Here, we look at some of the advantages and disadvantages of each model (Buckley, 1997; Davis, 2001, Aronoff, 1989).

1. The following are advantages of the raster data model:
   - Rasters are easy to understand, easy to read and write, and easy to draw on the screen.
   - It is a simple data structure.
   - The inherent nature and structure of raster data are ideal for mathematical modelling and quantitative analysis. Overlay operations are easily and efficiently implemented.
   - High spatial variability is efficiently represented in a raster format.
   - The raster format is more or less required for efficient manipulation and enhancement of digital images.
   - Remote sensing imagery is usually obtained in raster format and therefore is easily integrated.
   - In a raster model, discrete and continuous data types are accommodated equally and can be integrated easily.
2. The advantages of using vector data are the following:
   - The vector model can be used to represent data at its original resolution and shape without generalization.
   - Noncontinuous data (e.g., rivers, district boundaries, road lines, mountain peaks) is represented.
   - Vector data are very high resolution and therefore more accurate than raster data.
   - Vector displays and graphic outputs are pleasing to the eye as they more closely resemble hand-drawn maps.
   - Vector data provide efficient encoding of topology and, as a result, more efficient implementation of operations that require topological information, such as network analysis;
   - The vector data format takes less storage space and offers better storage capabilities.

3. Disadvantages of the raster data model are as follows:
   - The raster data structure is less compact.
   - Grids are poor at representing points, lines, and areas.
   - Because the cell size determines the resolution at which the data is presented, spatial inaccuracies are common in raster formats.
   - Topological relationships are more difficult to represent.
   - Grids suffer from the mixed-pixel problem.
   - Grids must often include redundant or missing data, leading to overly large files.
   - The output of graphics is less aesthetically pleasing because boundaries tend to have a blocky appearance rather than the smooth lines of hand-drawn maps.

4. The disadvantages of using vector data are as follows:
   - It is more difficult to manage vector data compared to raster data.
   - For effective analysis, vector data must be converted into a topological structure, and these analysis functions and technical aspects are complex.
   - Overlay operations are more difficult to implement.
   - Continuous data, such as elevation or precipitation data, are not well represented in vector form.
   - Manipulation and enhancement of digital images cannot be effectively done in the vector domain.
   - Vector data require machines that are more powerful and high tech.

4.10 Spatial Data Acquisition

Data acquisition or input of geospatial data in digital format is the most expensive part of setting up a GIS. On average, this forms about 80% of the total GIS project cost. Data acquisition and input are also time consuming and error prone. It is something you have to be careful about while setting up a GIS because the quality of the data obtained and the procedures used to input the data can have a profound effect on the products generated from the GIS. Some key issues that constrain the use of the GIS are data accessibility (ownership issues, fear of misuse, legal issues, protection, copyright issues); data costs (market demand drives pricing); data standards (conformity to give users confidence, credibility, acceptability, usefulness, and understanding); and metadata.

There are a number of sources for GIS data. Some of these are analog maps, aerial photographs, satellite images, ground surveys with a global positioning system (GPS), reports and publications, national mapping organizations, Census or Statistics Bureau, and clearinghouse (specialized digital data providers).

Once data has been obtained (either through fieldwork or from secondary sources such as maps and remotely sensed images), it has to be entered into the GIS. Data input is the procedure of encoding data into a computer-readable form and writing the data to the GIS database. The data input phase includes three main parts: data capture, editing and cleaning, and geocoding. The last two are also called data preprocessing.

The creation of an accurate and well-documented database is critical to the operation of a GIS. You must have heard the age-old saying, 'What you get out is what you put in'. There is a second one: GIGO or 'garbage in garbage out'. So, your GIS products are extensively dependent on the quality of the data you put into the system.

4.11 Data Quality Issues

The GIS analysis results, such as for combined maps, buffer zones, and other operations, can only be considered good and accurate if and only if the input data are of good quality. The dictionary term *quality* is defined as "degree of excellence" and indicates the goodness of the data (Heywood et al., 2002).

Data quality is a significant requirement in a GIS as it can have a significant impact on data conversion method selection and costs (Montgomery and Schuch, 1993). Data produced by mapping agencies and government departments often meet higher data standards as they assess the data quality to produce better results. Data created from different sources with different techniques can have discrepancies in terms of accuracy, precision, resolution, orientation, and displacements. An understanding of the concept of quality and the degree of quality is really important in digital data.

The following data quality information is of paramount importance: data quality information, date of collection, position accuracy, classification accuracy, and completeness. Data completeness is essentially a measure of totality of features. A complete or near-complete data set is one with a minimal amount of missing features. Position accuracy is the true representation of reality and the extent to which an estimated or measured data value approaches its true value (Aronoff, 1989). No data set can be 100% accurate; however, data can be accurate to within a specified tolerance (Heywood et al., 2002). Date of collection is important as it provides whether the data is still relevant. The data collected may be too old or may have been collected in a different season. Classification accuracy can be considered the discrepancy between the actual attribute value and coded attribute value.

Data quality concepts, standards, and benchmarks provide an important framework for data producers and users. A complete record of processes used to capture and manipulate data gives spatial data producers a better understanding of their databank and allows them to more efficiently manage data collection and distribution. This enables end users to use this information to determine the appropriateness and usefulness of a data set for a given application and reduces the possibility of misuse and abuse. The end users of the products generated will also have more confidence in the use of these products.

4.12 Sources of Error in GISs

Spatial and attribute errors can occur at any stage in a GIS. These errors can originate from a large number of different sources. The knowledge of these sources can reduce large amounts of errors in the data sets. The input data (source data) in a GIS may contain measurement inaccuracies or errors. For example, a mistake in operating systems and recording observations such as related to a GPS can result in errors in spatial data. Also, remotely sensed and aerial photography data with incorrect spatial reference and classification could create attribute errors. Data encoding, which is the process of transforming a non-GIS source into a GIS format, can lead to large errors in GISs. Digitizing, either manual or automatic, is another source of error. Errors can also occur after encoding and while data editing, cleaning, and conversion are undertaken. Errors may arise from data processing and analysis. GIS operations such as classification, aggregation, and integration can introduce errors. Therefore, the way the data are analysed and classified and how the data are interpreted plays an important role in minimizing the amount of errors. In addition, errors can be introduced in the GIS outputs and display. The extent of these inaccuracies is dependent on the attention paid during the construction, manipulation, and analysis phases (Heywood et al., 2002).

4.13 Issues of Scale

The technological innovations of GISs enable us to visualize the extensive geographical distributions of spatial features. GISs can enlarge (zoom in) or reduce (zoom out) data to any output scale, regardless of how sensible the results may be. Scale is a fundamental concept in GISs because it provides key information about the size, measurements, analysis, and modelling. Changing the scales in GIS does not require any changing process, generalization, or systematic aggregation of the elemental objects from the database (Longley and Batty, 1996). Combining data sets and overlaying different maps from different scales can cause confusion about the spatial accuracy of the output. Also, comparing maps with finer detail with maps of coarser detail can lead to invalid results in map comparisons (Zhang et al., 2014). Attention must also be paid to the scale incompatibility between GIS data and the original maps. A feature digitized from a 1:1,000,000 map sheet will not have the same precision and accuracy of a similar feature digitized from a 1:20,000 map sheet; therefore, the user must take care when making decisions based on the scale of the presented data.

4.14 Measurement Scales: Nominal, Ordinal, Interval, and Ratio

The level of measurement relates to the relationship between the values that are assigned to the attributes of a variable. There are four levels: nominal, ordinal, interval, and ratio. Knowledge of the level of measurement helps guide the decision on what statistical analysis can be applied and how to interpret the data from that variable. Not all statistical analyses are appropriate for all levels of data.

Nominal scales are used for labelling variables. They do not show any quantitative value; hence, nominal scales can simply be called labels. Nominal values or observations can be assigned a code in the form of a number, with the numbers simply labels. One can count but not order or measure nominal data. For example, gender and eye colour are nominal observations. You cannot order eye colour, as one colour is not necessarily better than another. Nominal variables are for mutually exclusive, but not ordered, categories.

Ordinal scales are one step above nominal scales in that they can be ordered. The order of the values is important, but the difference between the values has no meaning. Ordinal scales are typically measures of nonnumerical concepts such as satisfaction ratings. Most of us have undertaken a consumer survey to provide feedback on the level of service. So, the question could be, How satisfied were you with the level of service? and the selectable options would be 1 for very unsatisfied, 2 for somewhat unsatisfied, 3 for neutral, 4 for somewhat satisfied, and 5 for very satisfied. You can see that the order (ranking) is important here, but there is no meaning in subtracting the fifth category from the third category. The difference or distance between the categories does not have any meaning. One can count and order ordinal variables but cannot measure them. Compare this with nominal scales, which you can only count but cannot order or measure. Because ordinal data have no measure, you would not find an average of the values. One would normally use the mode or median as the measure of central tendency.

Interval scales are similar to ordinal scales except that the difference between values has a meaning or is interpretable. The intervals between values are equally split. Temperature measurement provides interval data. The difference between 30°C and 35°C is the same as the difference between 50°C and 55°C. Another example of interval scales is time. The increments are consistent and measurable. Because interval scales are numeric, this opens the option to use almost the full range of statistical analysis on these data sets. For example, central tendency can be measured using mode, median, mean, or standard deviation. However, it should be noted that in interval measurement ratios do not make any sense. As an example, 50°C is not twice as hot as 25°C even though the attribute value is twice as large. The idea of no temperature or no time does not exist. With interval data, we can add and subtract but cannot multiply or divide.

Ratio scales sit at the highest level in measurement scales. They give us order, an exact value between measurements, as well as an absolute zero. This allows one to apply the full range of descriptive and inferential statistics for use with such data. Ratio observations can be added, subtracted, multiplied, and divided. The full range of central tendency measures can be applied. Some examples of ratio observations are weight and height. A person who is 6 feet tall is twice as high as one who measures 3 feet, so a ratio makes sense as there is an absolute zero height. In epidemiological research, most 'count' variables are ratios. For example, counts of mosquitoes in a given region over a period of time are ratio scales because you can legitimately have zero counts of mosquitoes and it is also correct to say that you had twice the number of mosquitoes in January compared to June (ratio measurement). The same goes for the number of patients suffering from dengue fever who visit a hospital over a period of time.

4.15 Satellite Imagery as a Source of Spatial Data

The vectors and pathogens are able to survive at any given location because the environmental and climatological conditions at that location are conducive to their survival and population growth. The density, distribution, and survival of the vectors, and subsequently the pathogens, are determined to a large degree by the environment. Spatial and temporal changes in environmental conditions are important determinants of vector-borne disease transmission. Thus, the ability to map, assess, and monitor the environment and changes therein are critical for controlling vector-borne diseases. Satellite imagery has been capable of identifying these changes and supporting defining and predicting areas and periods of high transmission (Jeganathan et al., 2001; Kalluri et al., 2007; Khormi and Kumar, 2011). Satellite imagery has also been used for predicting temporal changes in infection rates (Khormi and Kumar, 2012; Leighton et al., 2012; Marechal et al., 2008). Associations between vector presence and density and satellite-image-derived variables such as humidity, temperature, land use, land cover, and rainfall have been used to identify, characterise, and map vector habitats (Nmor et al., 2013; Palaniyandi, 2014; Ra et al., 2012; Robinson et al., 1997a, 1997b; Rogers et al., 1996; Young et al., 2013). Thus, the potential for using remotely sensed images for monitoring and evaluating the factors associated with vector-borne diseases has been realized and is being used with increasing frequency.

Remote-sensing satellites provide continuous information on the environment and offer an opportunity for monitoring at scales and spatial resolutions that has not been available until recently. The past few decades have seen rapid growth in the number and capabilities of Earth observation satellites, which can be used for environmental mapping, monitoring, and assessment. Spatial resolutions have decreased from around 30 m to around 2 m in a decade. Although the Landsat series of satellites had 30-m resolutions at best, the current generation of satellites (e.g., Quickbird and Worldview) has spatial resolutions of around 2 m. Temporal resolution has also decreased, and for many parts of the world, daily images have become a reality.

Satellites can be used to capture data over large regions of the world where ground access is difficult and to monitor changes in natural resource distribution and variations in weather systems. They are especially useful in countries where infrastructure is less well developed, making fieldwork for data collection time consuming and expensive. Satellite imagery enables us to monitor large regions quickly and repeatedly, thus enabling us to detect important changes in the environment and the weather soon enough to take control and preventive action in relation to vector-borne diseases.

The paragraphs that follow summarise the use of spatial data derived from satellites in the mapping, monitoring, and modelling of vector-borne diseases.

4.15.1 Remote-Sensing Applications of Vector-Borne Diseases

Probably the most common use of satellite imagery in vector-borne diseases relates to mosquitoes. Mosquito-borne diseases are prevalent throughout the world and perhaps cause the largest number of deaths. The environmental and climatological conditions preferred by mosquitoes are reasonably well understood and are generally easy to map and monitor with satellite images. Hay et al. (2000) provided a summary of mosquito biology and methods to map their habitats using satellite data. In the early days of image data applications in epidemiology, the major focus was on identifying breeding habitats, such as marshes and wetlands (Barnes and Cibula, 1979; Kalluri et al., 2007; Wagner et al., 1979; Pope et al., 1992; Beck et al., 1994; Rejmankova et al., 1992). Wood et al. (1991a, 1991b) used a leaf area index (LAI) derived from Landsat data over 104 rice fields to compare with larval counts of *Aedes freeborni* from the edge of fields and close to livestock pastures to show that fields with a high LAI and near pastures had larger numbers of mosquitoes compared to fields that had a low LAI and were farther from pastures. Roberts et al. (1996) used Sattelite Pour l'Observation de la Terre (SPOT) imagery to develop a probability index of mosquito presence based on distance of houses from waterways, elevation above waterways, and forest cover between waterways and residences. Achee et al. (2006) used SPOT and IKONOS data to extract land cover information and relate this to larval habitats of *Anopheles darling*, reporting that there was a strong correlation between larval habitats and land cover types.

More recently, there has been a tendency to use data derived from satellite imagery together with meteorological variables to develop statistical models for mapping vector habitats. Linthicum et al. (1991) used the satellite image-derived Normalized Difference Vegetation Index (NDVI) with seasonal rainfall data to show variability in Rift Valley fever. Thomson et al. (1996) developed models incorporating NDVI and cold cloud duration (CCD) to map mosquito and malaria distribution in Africa; Rogers et al. (2002a) used NDVI, CCD, and land surface temperature to capture habitat seasonality of *Anopheles gambiae*. Linthicum et al. (1999) and Anyamba et al. (2001) showed that the incidence of Rift Valley fever in Eastern Africa could be forecast up to 5 months in advance by using sea-surface temperature variations and elevations from the Pacific and Indian Oceans together with NDVI data. Risk maps of West Nile virus for the United States and yellow fever and dengue for the whole world were created from satellite data and the entomological inoculation rate (EIR) (Rogers et al., 2002b, 2006).

Apart from the mapping of mosquito-borne diseases using satellite imagery, many other applications exist for other vectors. Hugh-Jones (1991a, 1991b) derived land cover maps from Landsat images to discriminate four grazing regions with different levels of cattle tick *Amblyomma variegatum* infestation. Randolph (1994, 2000) correlated multitemporal NDVI with mean mortality rates of cattle to investigate seasonal-dependent mortality rates caused by ticks. Glass et al. (1995) showed that many suitable tick habitats in the United States correlated with residential properties close to wooded areas. Landsat data was used to extract brightness, greenness, and wetness indices. Randolph and Rogers (2006) used environmental data derived from satellite images to differentiate the ecoclimatic zones of six tick-borne flaviviruses, concluding that climate may have played a role in directing their evolution.

Thomson et al. (2000) used forest cover and land cover classes derived from the Advanced Very High Resolution Radiometer (AVHRR), together with topography and soil data, to predict the prevalence of *Loa loa* in six African countries. About 50% of the variation in the prevalence rates of infection could be explained by environmental variables derived from satellite imagery. Rogers (1979) used air temperature and vapour pressure derived from satellite data to successfully model the distribution and density of tsetse flies over Africa; Rogers et al. (1996), Rogers (2000), and Green and Hay (2002) used NDVI, land surface temperature, and cold cloud temperature duration to model tsetse distribution over Africa. Cross et al. (1996) used NDVI and meteorological data to map the distribution of *Phlebotomus papatasi* in the Middle East. Land cover data derived from the Indian Remote Sensing Satellite Linear Imaging Self-Scanning System 3 (IRS LISS3) were compared with sandfly (*Phlebotomus argentipes*) densities from different areas, and it was noted that endemic areas had a higher percentage of water bodies and marshes.

These are only a few examples of the use of satellite data for mapping vectors and vector-borne diseases. The opportunities for using satellite data for mapping and modelling of vectors are endless. The opportunities will increase with the increasing quality of satellite data. Frequency of data capture, spatial resolutions, and available satellites is literally increasing by the day. It is up to researchers in this field to develop new applications and think of novel ways in which this plethora of underutilized data can be used to support the decision-making process.

4.16 Conclusion

Spatial data is a critical component of any GIS system. No matter how powerful or expensive the system is, without spatial data the system is generally useless. There are many sources of spatial data. Some of these sources make spatial data freely available; others have costs involved. In spatial analysis, one normally uses a mix of self-collected data and data obtained from other sources. Environmental data over large areas is generally obtained from vendors supplying satellite data; socioeconomic and demographic data is mostly obtained from government organizations. More specific data pertaining to particular projects and especially those over small regions is self-collected. Self-collected data is normally more accurate, especially if the data are point data collected using systems such as a differential GPS.

In epidemiological studies, disease incidence data or data pertaining to health services is not readily available or easily accessible; much of this has to do with privacy issues. Thus, spatial analysis of disease incidences and linking them with environmental and climatological variables are difficult. Where incidence data is available, this is often aggregated so patient identities are not revealed. This results in spatial analysis undertaken with aggregated data, so fine-scale intercorrelations with other variables becomes impossible.

The quality of spatial data is also critical for determining causal effects. It is a paradox that although disease incidence data is generally the least readily available or accessible, these data are the most accurate. Addresses of patients are recorded by house numbers, more so in developed countries compared to developing countries, so they can be located on a map to within a few metres. Satellite data accuracies have improved over the years. There are better image rectification and geo-referencing procedures available now compared to a decade ago. The spatial resolution of images has also improved considerably, with submetre images available from a variety of satellites. These high-resolution images permit fine-scale mapping of environmental variables and thus enable modelling to be undertaken using disease incidence data at the point scale rather than being aggregated to district scales. These developments have enabled higher accuracy and simulation models that are more reliable to be developed, as will be shown in further chapters.

References

Achee, N.L., Grieco, J.P., Masuoka, P., Andre, R.G., Roberts, D.R., Thomas, J., Briceno, I., King, R., Rejmankova, E. (2006). Use of remote sensing and geographic information systems to predict locations of *Anopheles darlingi*-positive breeding sites within the Sibun River in Belize Central America. *Journal of Medical Entomology,* 43: 382–392.

Altibase. (2014). *Altibase Application Development Spatial SQL Reference Release 6.3.1.* Seoul, Korea: Altibase Corporation.

Anyamba, A., Linthicum, K.J., Tucker, C.J. (2001). Climate-disease connections: Rift Valley fever in Kenya. *Cadernos de Saude Publica,* 17: 133–140.

Aronoff, S. (1989). *Geographic Information Systems: A Management Perspective.* Ottawa, Canada: WDL.

Barnes, C.M., Cibula, C.G. (1979). Some implications of remote sensing technology in insect control programs including mosquitoes. *Mosquito News,* 39: 271–282.

Beale, L., Abellan, J.J., Hodgson, S., Jarup, L. (2008). Methodologic issues and approaches to spatial epidemiology. *Environmental Health Perspectives,* 116, 1105–1110.

Beck, L.R., Rodriguez, M.H., Dister, S.W., Rodriguez, A.D., Rejmankova, E., Ulloa, A., Meza, R.A., Roberts, D.R., Paris, J.F., Spanner, M.A., Washino, R.K., Hacker, C., Legters, L.J. (1994). Remote sensing as a landscape epidemiologic tool to identify villages at high risk for malaria transmission. *American Journal of Tropical Medicine and Hygiene,* 51: 271–280.

Bernhardsen, T. (2002). *Geographic Information Systems: An Introduction.* Arendal, Norway: Wiley.

Blan, L., Butler, R. (1999). Comparing effects of aggregation methods on statistical and spatial properties of simulated spatial data. *American Society for Photogrammetry and Remote Sensing,* 65, 73–84.

Buckley, D.J. (1997). *The GIS Primer: An Introduction to GIS.* Fort Collins, CO: Pacific Meridian Resource.

Burrough, P.A. (1986). *Principles of Geographic Information Systems for Land Resource Assessment.* Oxford, UK: Oxford University Press.

Chang, K.-T. (2002). *Introduction to Geographic Information Systems.* Boston: McGraw-Hill.

Chrisman, N.R. (1997). *Exploring Geographic Information Systems.* New York: Wiley.

Cross, E.R., Newcomb, W.W., Tucker, C.J. (1996). Use of weather data and remote sensing to predict the seasonal distribution of *Phlebotomus papatasi* in southwestern Asia. *American Journal of Tropical Medicine and Hygiene,* 54: 530–536.

Cui, T. (2001). Characteristic of spatial data and the design of data model. Paper presented at the 20th International Cartographic Conference, Beijing, China, August 6–10.

Davis, B.E. (2001). *GIS: A Visual Approach.* Albany, NY: Onword Press Thomson Learning.

Dent, B.D. (1999). *Cartography: Thematic Map Design.* New York: McGraw-Hill.

Glass, G.E., Schwartz, B.S., Morgan, J.M., III, Johnson, D.T., Noy, P.M., Israel, E. (1995). Environmental risk factors for Lyme disease identified with geographic information systems. *American Journal of Public Health,* 85: 944–948.

Green, R.M., Hay, S.I. (2002). The potential of Pathfinder AVHRR data for providing surrogate climatic variables across Africa and Europe for epidemiological applications. *Remote Sensing of Environment,* 79: 166–175.

Hay, S.I., Omumbo, J.A., Kraig, M.H., Snow, R.W. (2000). Earth observation geographic information systems and *Plasmodium falciparum* malaria in sub-Saharan Africa. *Advances in Parasitology,* 47: 174–206.

Heywood, I., Cornelius, S., Carver, S. (2002). *An Introduction to Geographical Information Systems.* Harlow, UK: Pearson Education.

Hugh-Jones, M. (1991a). Landsat-TM identification of the habitats of the cattle tick *Amblyomma variegatum* in Guadeloupe French Windward Islands. *Preventive Veterinary Medicine,* 11: 355–356.

Hugh-Jones, M. (1991b). The remote sensing of tick habitats. *Journal of Agricultural Entomology,* 8: 309–315.

Johnson, V. (2011). I object to your projections. In: Herrman, J., Buchanan, M. (Eds.), *The Awl.* http://www.theawl.com/2011/09/i-object-to-your-projections

Jeganathan, C., Khan, S.A., Chandra, R., Singh, H., Srivastava, V., Raju, P.L.N. (2001). Characterisation of malaria vector habitats using remote sensing and GIS. *Journal of the Indian Society of Remote Sensing,* 29(1): 31–36.

Kalluri, S., Gilruth, P., Rogers, D., Szczur, M. (2007). Surveillance of arthropod vector-borne infectious diseases using remote sensing techniques: a review. *PLoS Pathogens*, 3(10): e116.

Kemp, K. (2008). *Encyclopedia of Geographic Information Science*. Thousand Oaks, CA: Sage.

Khormi, H.M., Kumar, L. (2011). Identifying and visualizing spatial patterns and hot spots of clinically-confirmed dengue fever cases and female *Aedes aegypti* mosquitoes in Jeddah, Saudi Arabia. *Dengue Bulletin*, 35: 15–34.

Khormi, H.M., Kumar, L. (2012). Assessing the risk for dengue fever based on socioeconomic and environmental variables in a geographical information system environment. *Geospatial Health*, 6: 171–176.

Leighton, P.A., Koffi, J.K., Pelcat, Y., Lindsay, L.R., Ogden, N.H. (2012). Predicting the speed of tick invasion: an empirical model of range expansion for the Lyme disease vector *Ixodes scapularis* in Canada. *Journal of Applied Ecology*, 49: 457–464.

Linthicum, K.J., Anyamba, A., Tucker, C.J., Kelley, P.W., Myers, M.F., Peters, C.J. (1999). Climate and satellite indicators to forecast Rift Valley fever epidemics in Kenya. *Science*, 285: 397–4000.

Linthicum, K.J., Bailey, C.L., Tucker, C.J., Angleberger, D.R., Cannon, T., Logan, T.M., Gibbs, P.H., Nickeson, J. (1991). Towards real-time prediction of Rift-Valley fever epidemics in Africa. *Preventive Veterinary Medicine*, 11: 325–334.

Longley, P.A., Batty, M. (1996). *Spatial Analysis: Modelling in a GIS Environment*. New York: Wiley.

Marechal, F., Ribeiro, N., Lafaye, M., Güell, A. (2008). Satellite imaging and vector-borne diseases: the approach of the French National Space Agency (CNES). *Geospatial Health*, 3(1): 1–5.

McDougall, K., Rajabifard, A., Williamson, I.P. (2007). A mixed method approach for evaluating spatial data sharing partnerships for Spatial Data Infrastructure development. In Onsrud, H. (Ed.), *Research and Theory in Advancing Spatial Data Infrastructure Concepts*. Redlands, CA: ESRI Press.

Montgomery, G.E., Schuch, H.C. (1993). *GIS Data Conversion Handbook*. Fort Collins, CO: GIS World and UGC Consulting.

Nmor, J.C., Sunahara, T., Goto, K., Futami, K., Sonye, G., Akweywa, P., Dida, G. Minakawa, N. (2013). Topographic models for predicting malaria vector breeding habitats: potential tools for vector control managers. *Parasites and Vectors*, 6: 14. doi:10.1186/1756-3305-6-14

Onsrud, H.J., Rushton, G. (1995). Sharing geographic information: an introduction. In: Onsrud, H.J., Rushton, G. (Eds.), *Sharing Geographic Information*. New Brunswick, NJ: Centre for Urban Policy Research.

Palaniyandi, M. (2014). GIS for disease surveillance and health information management in India. *Geospatial Today*, 13(5): 44–46.

Pope, K.O., Sheffner, E.J., Linthicum, K.J., Bailey, C.L., Logan, T.M., Kasischke, E.S., Birney, K., Njogu, A.R., Roberts, C.R. (1992). Identification of central Kenyan Rift Valley fever virus vector habitats with Landsat TM and evaluation of their flooding status with airborne imaging radar. *Remote Sensing of Environment*, 40: 185–196.

Ra, P.K., Nathawat, M.S., Onagh, M. (2012). Application of multiple linear regression model through GIS and remote sensing for malaria mapping in Varanasi District, India. *Health Science Journal*, 6(4): 731–749.

Randolph, S.E. (1994). Population dynamics and density-dependent seasonal mortality indices of the tick *Rhipicephalus appendiculatus* in eastern and southern Africa. *Medical and Veterinary Entomology*, 8: 351–368.

Randolph, S.E. (2000). Ticks and tick-borne disease systems in space and from space. *Advances in Parasitology*, 47: 217–240.

Randolph, S.E., Rogers, D.J. (2006). Tick-borne disease systems: mapping geographic and phylogenetic space. *Advances in Parasitology*, 62: 263–291.

Rao, A.S. (2012). What do you mean by GIS aggregation? http://www.publishyourarticles.net/knowledge-hub/geography/what-do-you-mean-by-gis-aggregation.html

Rejmankova, E., Savage, H.M., Rodriguez, M.H., Roberts, D.R., Rejmanek, M. (1992). Aquatic vegetation as a basis for classification of *Anopheles albimanus* Wiedemann (Diptera: Culcidae) larval habitats. *Environmental Entomology*, 21: 598–603.

Roberts, D.R., Paris, J.F., Manguin, S., Harbach, R.E., Woodruff, R., Rejmankova, E., Polanco, J., Wullschleger, B., Legters, L.J. (1996). Predictions of malaria vector distribution in Belize based on multispectral satellite data. *American Journal of Tropical Medicine and Hygiene*, 57: 304–308.

Robinson, T., Rogers, D.J., Williams, B. (1997a). Mapping tsetse habitat suitability in the common fly belt of southern Africa using multivariate analysis of climate and remotely sensed vegetation data. *Medical and Veterinary Entomology*, 11: 235–245.

Robinson, T.P., Rogers, D.J., Williams, B. (1997b). Univariate analysis of tsetse habitat in the common fly belt of southern Africa using climate and remotely sensed vegetation data. *Medical and Veterinary Entomology*, 11: 223–234.

Rogers, D. (1979). Tsetse population dynamics and distribution: a new analytical approach. *Journal of Animal Ecology*, 48: 825–849.

Rogers, D.J. (2000). Satellites space time and the African trypanosomiasis. *Advances in Parasitology*, 47: 129–171.

Rogers, D.J., Hay, S.I., Packer, M.J. (1996). Predicting the distribution of tsetse flies in West Africa using Fourier processed meteorological satellite data. *Annals of Tropical Medicine and Parasitology*, 90: 225–241.

Rogers, D.J., Myers, M.F., Tucker, C.J., Smith, P.F., White, D.J., et al. (2002a). Predicting the distribution of West Nile fever in North America using satellite sensor data. *Photogrammetric Engineering and Remote Sensing*, 68: 112–114.

Rogers, D.L., Randolph, S.E., Snow, R.W., Hay, S.I. (2002b). Satellite imagery in the study and forecast of malaria. *Nature*. 415: 710–715.

Rogers, D.J., Wilson, A.J., Hay, S.I., Graham, A.J. (2006). The global distribution of yellow fever and dengue. *Advances in Parasitology*, 62: 181–220.

Sahu, K.C. (2007). *Textbook of Remote Sensing and Geographical Information Systems*. New Delhi, India: Atlantic.

Thomson, M.C., Connor, S.J., Milligan, P.J.M., Flasse, S.P. (1996). The ecology of malaria seen by earth-observation satellites. *Annals of Tropical Medicine and Parasitology*, 90: 243–264.

Thomson, M.C., Obsomer, V., Dunne, M., Connor, S.J., Molyneux, D.H. (2000). Satellite mapping of *Loa loa* prevalence in relation to ivermectin use in west and central Africa. *Lancet*, 356: 1077–1078.

Wagner, V.E., Hill-Rowley, R., Newson, H.D. (1979). Remote sensing: a rapid and accurate method of data acquisition for a newly formed mosquito control district. *Mosquito News*, 39: 271–282.

Williamson, I.P., Rajabifard, A., Feeney, M.-E.F. (2003). *Developing Spatial Data Infrastructures: From Concept to Reality*. London: Taylor and Francis.

Wood, B.L., Beck, L.R., Washino, R.K., Palchick, S.M., Sebesta, D. (1991a). Spectral and spatial characterization of rice field mosquito habitat. *International Journal of Remote Sensing*, 12: 621–626.

Wood, B., Washino, R., Beck, L., Hibbard, K., Pitcairn, M., Roberts, D., Rejmankova, E., Paris, J., Hacker, C., Salute, J., Sebesta, P., Legters, L. (1991b). Distinguishing high and low *Anopheline*-producing rice fields using remote sensing and GIS technologies. *Preventive Veterinary Medicine*, 11: 277–288.

Young, S.G., Tullis, J.A., Cothren, J. (2013). A remote sensing and GIS assisted landscape epidemiology approach to West Nile virus. *Applied Geography*, 45: 241–249.

Zhang, J., Atkinson, P., Goodchild, M.F. (2014). *Scale in Spatial Information and Analysis*. Boca Raton, FL: Taylor & Francis.

Chapter 5

Common Spatial Methods for Modelling and Analysing Spatial and Temporal Patterns and Distributions of Mosquito- Borne Diseases

5.1 Introduction

The first step in understanding diseases is to identify patterns. Data for disease occurrence and outbreak is usually characterised by spatial structures because of spatial autocorrelation.

Despite advances in medical knowledge, the etiological and pathological processes of many diseases still need to be better understood. Epidemiological knowledge is often statistical, based on the relation between occurrence rates and the expected factors. One piece of the puzzle that needs to be solved is the spatial characteristics of the disease. Spatial variations are believed to be well known, and epidemiology factors in as a critical aspect.

Much of the interest in spatial data analytics systems is based on solving the problems in plant or animal ecology and agricultural field trials and its applications. But, the application of these methods for the analysis of disease patterns is not always obvious to researchers. However, there has been growing interest in this field, mainly in response to vector-borne diseases, particularly ones that have a high impact on society, such as dengue fever (DF). This interest is further fuelled by its need because of the spatiotemporal nature of DF.

The use of spatial statistics to find such patterns provides a better understanding of the geographic nature of DF and other vector-borne diseases by allowing researchers to pinpoint the causes of specific geographical patterns. Thus, the decisions can be made with a higher level of confidence, and summarising the patterns through a comprehensive set of details.

Geographic information systems (GISs) have a proven potential for addressing epidemiological problems. The use of spatial analysis tools is important to identify critical control areas with several variables intimately related to the modulation of the disease dynamics. Therefore, a GIS is a valuable tool for investigating patterns in a DF outbreak.

Methods have been developed to describe spatial patterns of point locations or events. Events are any sort of phenomena that occur at specific locations at a specific time, such as locations of disease outbreaks or locations of mosquito egg-laying sites.

Identifying the pattern of events is the first step in understanding the process responsible for that pattern. Events may be dispersed from one another, clustered together, or occur at random. For example, a disease, such as DF, may occur in spatially clustered locations, suggesting a localised environmental factor may be responsible. Linking pattern to process is often the purpose of identifying spatial patterns of events.

5.2 Pattern Analysis

5.2.1 Average Nearest Neighbour

Point pattern analysis, in its basic form, deals with the distribution of homogeneous points—that is, one type of point. This does not imply that we cannot treat points that are not homogeneous in the real world. In basic point pattern analysis, we focus only on the spatial aspect of point distributions, neglecting their attributes. Here, with respect to vector-borne diseases such as DF, borne by the *Aedes* mosquito, we can consider the points to be case occurrences or the *Aedes* nesting locations (Tipayamongkholgul and Sunisa, 2011).

In point pattern analysis, it is important to consider whether a point distribution shows a clustered pattern or a dispersed pattern.

If points represent cases of a disease, a point cluster suggests that the disease is epidemic or that there is a source of, for example, a water hole near the cluster (Figure 5.1).

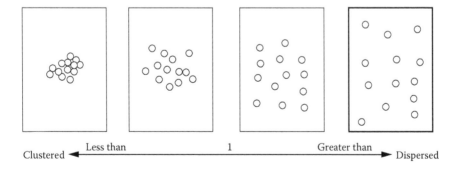

Figure 5.1   Point distribution shows a clustered pattern or dispersed pattern.

In point pattern analysis, we use a quantitative measure that indicates the degree of clustering. To describe the degree of spatial clustering of a point distribution, the nearest-neighbour distance method uses the average distance from every point to its nearest-neighbour point. The standardised nearest-neighbour distance is a descriptive measure of the degree of point clustering. Indexes of less than 1 mean the patterns are clustered, and indexes greater than 1 mean the patterns are dispersed (Figure 5.1). The average nearest neighbour (ANN) is as follows:

$$\text{ANN} = \frac{\text{Observed average distance (OAD)}}{\text{Expected average distance (EAD)}}$$

$$\text{OAD} = \frac{\sum_{i=1}^{n} d_i}{n}$$

$$\text{EAD} = \frac{0.5}{\sqrt{n/A}}$$

where
  d_i = distance to the nearest point from point i
  n = number of points
  A = the area of a minimum enclosing rectangle around all features

The nearest-neighbour distance thus defined is an 'absolute' measure of point clusters. It depends on the size of the region in which points are distributed, so we cannot compare two sets of points distributed in regions of different sizes. This indicates that the concept of a spatial cluster is based on the pattern of points with respect to the size of the region in which the points are located. To evaluate the degree of spatial clustering, therefore, we have to standardise the nearest-neighbour distance, taking the region size into account. As an

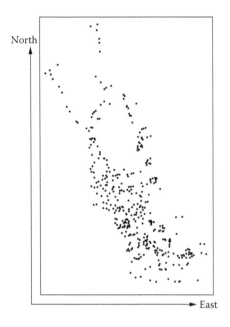

Figure 5.2   The patterns of DF cases are clustered according to the ANN index.

example, an average nearest-neighbour distance analysis can be performed to measure the distance between each dengue case and its nearest neighbour's location. All these average distances are then taken and divided by the expected average distance to generate an index of 0.76, suggesting the patterns are clustered and unlikely random (Figure 5.2).

5.2.2 *Getis-Ord Gi**

In the study of vector-borne diseases such as DF, methods that provide clustering are of great importance in helping to identify endemic areas and hot spots. Temporal analysis of DF has also been largely focused on clustering in other studies (Bartlett, 1947; Khormi et al., 2011; Lay et al., 2006; Wen et al., 2006; Wu et al., 2009). This provides a general risk factor across areas that are observed to be affected and reveals the nature of the DF clusters (Bithell, 1999).

Previous studies conducted on the biological vector of DF, *Aedes aegypti*, have found its larval density to be a good factor to consider, particularly when studying the clustering nature of DF in Rio de Janeiro, Brazil (Lagrotta et al., 2008).

Generally, DF clustering studies have not focused on the temporal aspect in spatial analysis (Earnest et al., 2005; Zeger et al., 2006). Often, ignorance of the temporal aspect results in difficulty in the decision-making process on any actionable plan derived from the study or even difficulty in simply assessing the degree of the DF endemic outbreak and how past measures have had an impact on it.

A good example of spatiotemporal analysis is the Getis-Ord Gi* statistic. It is a type of analysis particularly helpful for resource allocation and decision making and can be combined with a frequency index. These have been used together before for mapping, for example, DF temporal risk characteristics and modelling risk changes (Khormi et al., 2011).

Getis-Ord Gi* statistics identify different spatial clustering patterns like hot spots, high-risk areas, and cold spots over the entire study area, with statistical significance based on risk factors (Songchitruksa et al., 2010). These spots signify the risk of vector-borne diseases in these areas.

We start by analysing each district with respect to the frequency of transmission per unit time, and a frequency index f is utilised. This is calculated, and different levels of granularity of time (weekly, monthly, and then yearly) are calculated for each district. On a daily basis, it is based on the number of cases recorded such that for any number of occurrences, the value for that day would be 1, whereas for no occurrences, the value of that particular day is taken to be 0 (Giesecke, 2001; Wen et al., 2006).

The index is defined as ED/TD. Here, ED is equivalent to the cumulative number of days when at least one case of DF was recorded, and TD is the total number of days during the period (TD). For example, if over 1 week DF cases were reported on 2 days, the occurrence would be 2, resulting in a frequency index of 2/7.

The frequency index taken at different units of time is represented as AWFI for average weekly, AMFI for average monthly, and AYFI for an average annual basis. As the range of the frequency index value starts from 0 and ends at 1, the values that are tending toward 1 point toward a higher possible risk of DF; on the other hand, values tending toward 0 represent less likeliness of DF occurrence (Khormi et al., 2011).

Similarly, besides using DF cases as the base metric, the number of females of the DF vector *A. aegypti* can also be used in the form of their count derived from the readings of black hole traps. If the number of black hole traps differs between different districts, the number of female specimens collected from each trap must be divided by the number of traps for each district, resulting in the average number of vectors per trap.

One of the frequency indices can then be analysed using the Getis-Ord Gi* statistic to model the risk levels in each district on a unit time basis. This method factors in the context based on neighbouring values; specifically, if a certain feature value is significantly high where its neighbouring features also exhibit a high value, then we conclude that they collectively form a hot spot (this is mathematically detailed in Equation 5.2).

As an example, with any unit of time (i.e. 2010 months), the Getis-Ord equation can be used to calculate the spatio-temporal risk of DF to model the risk factor of each district. Districts with DF frequency index values can be cross-checked with observable high numbers of cases that are geographically homogeneous, thus revealing the districts with the highest risk of DF (Figure 5.3).

After that, the first step is to conceptualise the spatial relationships relating case locations in districts that are considered utilizing the fixed-distance band, which allows factoring in the

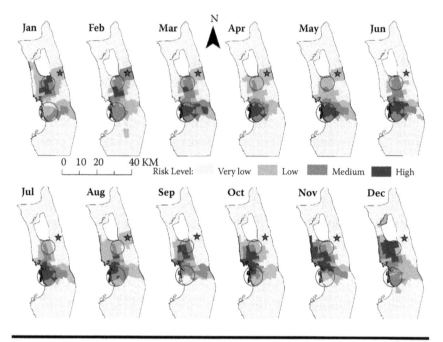

Figure 5.3   Spatiotemporal risk changes over 12 months in Jeddah districts.

frequency index of DF cases inside the boundary of the study area but excludes unrelated data that exists outside the boundary. This approach was chosen as it has been shown to be generally more appropriate than inverse distance conceptualisation (Mitchell, 2005; Khormi et al., 2012).

The second step is to get the z-score value and p value of each district under study utilizing the Euclidean distance.

Areas with high z scores and small p values exhibit the spatial clustering of a high level of DF hot spots (a high temporal risk in a given period), whereas areas with low z scores and high p values indicate a spatial clustering of a low level of DF hot spots (a low temporal risk in a given period).

The areas can be classified based on z-score values, for example: z scores of 3 or greater indicate high-risk areas, z scores of 2–3 indicate medium-risk areas, z scores of 1–2 indicate low-risk areas, and z scores of 1 or less indicate very-low-risk areas (Figure 5.3).

$$ZG_i^* = \frac{\sum_{j=i}^{n} w_{ij}x_j - x \times \sum_{j=i}^{n} w_{ij}^2}{s\sqrt{\dfrac{n\sum_{j=i}^{n} w_{ij}^2 - \left(\sum_{j=i}^{n} w_{ij}\right)^2}{n-1}}} \tag{5.2}$$

where

x_j = the attribute value for feature j
$w_{i,j}$ = the spatial weight between i and j
n = the total number of features

The results are presented on a monthly basis from January 2010 to December 2010 in Figure 5.3. They indicate hot spot areas with significantly high frequency index findings. In central Jeddah, the study area, the spatiotemporal monthly hot spots shown were distributed; see the larger circle and the Alsafa area (marked as the smaller circle). In addition, the results showed that the pattern of risk changed with time; for example, the Burman district (marked as a star) was identified as a low-risk area in January, March, April, May, and August; a medium-risk area in February; and a very-low-risk area in June, July, September, October, November, and December.

This method indicates that epidemiologists can identify disease case clusters when they factor in temporal properties. The spatiotemporal risk details presented in the example confirm that the temporal risk model—based on daily, weekly, or monthly frequency indices—produces better understanding of the changes compared with previous models based on a year. This provides insights for improving the vector-borne disease surveillance system and for controlling interventions in any affected area.

5.2.3 Ripley's K Function

Ripley's $K(t)$ function is for analysing completely mapped spatial point process data—that is, data on the locations of events.

These are usually recorded in two dimensions, but they may be locations along a line or in space (Dixon, 2006). Here, we focus on $K(t)$ for one-dimensional data. Ripley's $K(t)$ function can be used to summarise a point pattern, test hypotheses about the pattern, estimate parameters, and fit models. Bivariate or multivariate generalisations can be applied for explaining relationships between two or more point patterns. The multidistance spatial cluster (Ripley's K function) is used to estimate the spatial pattern and distribution (Khormi and Kumar, 2011b).

This method is useful for point pattern analysis, and it is also the best method to illustrate the point pattern at multiple distances compared with similar methods. For example, Ripley's K function can be used to determine whether the distribution of the clinically confirmed dengue cases of 2010 were clustered or dispersed at multiple distances (Figure 5.4). The inputs of value for this analysis are based on data from individual events, for example, case locations. The outputs can be represented as graphic models for the duration of epidemics (Figure 5.4).

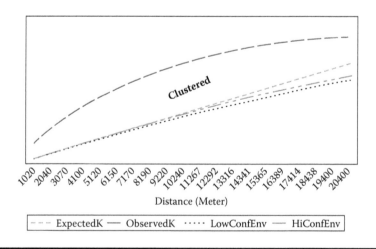

Figure 5.4 Spatial patterns of DF cases at different multiple distances.

The following K function can be used for analysis:

$$L(d) = \sqrt{\frac{A \sum_{i=1}^{n} \sum_{j=1,\ j!=i}^{n} k(i,j)}{\pi n(n-1)}} \qquad (5.3)$$

where

d = the distance

n = the total number of clinically confirmed DF cases

A = the total study area

$k_{i,j}$ = a weight, which (if there is no edge correction) is 1 when the distance between i and j is less than or equal to d and 0 when the distance between i and j is greater than d. When edge correction is applied, the weight of $k(i,j)$ is modified slightly

The K function is

$$K(t) = \lambda^{-1} E$$

where E is the number of extra events within distance t of a randomly chosen event.

Here, λ is the density of events (number per unit area). It can be estimated as

$$\lambda = N/A$$

where

N = the observed number of points

A = the interval of the study region

If edge effects are ignored, then $K(t)$ describes characteristics of the point processes at many distance scales. $K(t)$ is not uniquely determined by the point processes, so two different point processes could have the same $K(t)$ function.

The neighbourhood sizes and analysis of the data sets of clinically confirmed DF cases were put through 20 iterations for clustering to optimise accuracy. The confidence envelopes were computed at a 99% confidence interval, and weight fields were selected according to the layer data sets. For example, the number of cases in each district was used as a weight field when the method was used to analyse the spatial pattern of confirmed cases of DF. Simulated outer boundary values were selected as a boundary correction method because they simulated points outside the study area and because the simulated points mirrored points across the study area boundary.

Figure 5.4 shows the results of this analysis, which can be interpreted as follows: If the observed K value is larger than the expected K value for a particular distance, the distribution is more clustered than a random distribution at that distance (scale of analysis). If the observed K value is smaller than the expected K, the distribution is more dispersed than a random distribution at that distance. Also, if the observed K value is larger than the high-confidence envelope (HiConfEnv) value, spatial clustering for that distance is statistically significant. If the observed K value is smaller than the low-confidence envelope (LwConfEnv) value, spatial dispersion for that distance is statistically significant (Boots and Getis, 1988; Khormi and Kumar, 2011b).

In general, Figure 5.4 and Table 5.1 show that distributions of DF cases were clustered at multiple distances because the observed K values were larger than the expected K values at different distances with statistically significant clustering. The observed K value (minimum distance ≈ 4241 m and maximum distance ≈ 26,010 m) was larger than the expected K value (minimum distance ≈ 1020 m and maximum distance ≈ 20,400 m). As a result, the distribution of DF cases in this year was more clustered than a random distribution at those distances. Also, the mean distance of the observed K (≈ 18,907 m) was larger than the mean distance of high confidence (≈ 10,159 m), which confirmed that the spatial clustering at different multiple distances was statistically significant (Table 5.1 and Figure 5.4).

Table 5.1   Results of Replay's *K* Function Analysis

	LowConfEnv	HiConfEnv	ExpectedK	ObservedK
1	1008	1098	1020	4241
2	2023	2174	2040	7100
3	3032	3205	3070	9521
4	4032	4244	4100	11761
5	5000	5255	5120	13654
6	5948	6265	6150	15311
7	6874	7252	7170	16893
8	7803	8228	8190	18273
9	8689	9177	9220	19465
10	9559	10103	10240	20512
11	10407	11004	11267	21452
12	11227	11878	12292	22256
13	12029	12728	13316	22957
14	12808	13559	14341	23596
15	13568	14346	15365	24139
16	14300	15111	16389	24631
17	15007	15850	17414	25093
18	15690	16571	18438	25483
19	16348	17243	19400	25793
20	16975	17890	20400	26010

5.2.4 Moran's I

Moran's *I* measures the spatial correlation of a disease based on the disease locations and the disease values simultaneously. It assesses whether the disease pattern expressed is dispersed, clustered, or random. The Moran's *I* index value is calculated using this method, and both a *p* value and a *z* score evaluate the significance of that index. The value of spatial autocorrelation near −1 indicates dispersion and near +1 indicates clustering (Khormi and Kumar, 2011a). The null hypothesis indicates no spatial clustering of the values associated with geographic features in the study; as a result, the null hypothesis can be rejected (Upton and Fingleton, 1985; Getis and Ord, 1992). Moran's *I* has been used in many studies of vector-borne diseases. For example, Kitron and Kazmierczak (1997) used it to identify the degree of spatial clustering of Lyme disease cases, ticks, and forested vegetation, and they also used it to determine the distances where spatial effects are maximised. As a result, they assigned cases to the county level, and the spatial pattern that they found indicated relatively large geographic areas (clusters of counties) as risk areas for Lyme disease. Nakhapakorn and Jirakajohnkool (2006) mapped the local indicators of spatial autocorrelation in Sukhothai Province, Thailand. In this study, the local Moran's *I* reflected that the average of DF and dengue haemorrhagic fever (DHF) case prevalence in the northern part of the province had the highest local Moran's value. Results of this type of analysis lead to more comprehensive knowledge of how spatial patterns will change from the past to the future.

Consider the observations and locations where we define y_i as the number of cases and n_i as the population at risk at geographic unit *I*, where $i = 1, ..., N$ with *N* the total number of geographic units (e.g., census tracts or counties). Let w_{ij} be the weight assigned to the pair of geographic units *i* and *j* $(i \neq j)$, which reflects the strength of the relationship between geographic units *i* and *j*.

Moran's *I* is defined as

$$I = \left(\frac{1}{s_y^2} \right) \frac{\sum\limits_{i}^{N} \sum\limits_{\{j:\, i \neq j\}}^{N} w_{ij} \left(y_i - \bar{y} \right) \left(y_j - \bar{y} \right)}{\sum\limits_{i}^{N} \sum\limits_{\{j:\, i \neq j\}}^{N} w_{ij}} \tag{5.4}$$

where

$$\bar{y} = \sum_i y_{i/N}, \quad s_y^2 = \frac{1}{N} \sum_{i=1}^{N} (y_i - \bar{y})^2.$$

We note that y_i are counts; however, alternative versions of Moran's I use continuous values. The weight in Equation 5.4 is commonly defined based on adjacent neighbours (Adj) and is written as

$$w_{ij} = \begin{cases} 1 & \text{if } i, j \text{ are adjacent neighbours} \\ 0 & \text{otherwise.} \end{cases}$$

The weight w_{ij} in Moran's I and its extensions are usually defined as in Equation 5.2 (neighbour matrix). However, the weight function w_{ij} can be defined in many other ways.

$$S_{yy,w} = \frac{\sum_i^N \sum_{\{j:\, i \neq j\}}^N w_{ij} (y_i - \bar{y})(y_j - \bar{y})}{\sum_i^N \sum_{\{j:\, i \neq j\}}^N w_{ij}}$$

Then, we can rewrite Moran's I as

$$I = \frac{S_{yy,w}}{s_y^2}.$$

The value of I usually ranges between −1 and 1, and the expected value is $E[I] = -1/N - 1$. However, an example was shown to provide an overview of mosquito patterns in Jeddah (Figure 5.5). As mentioned, the range of I depends on the values of the weight function. The resulting figure shows positive

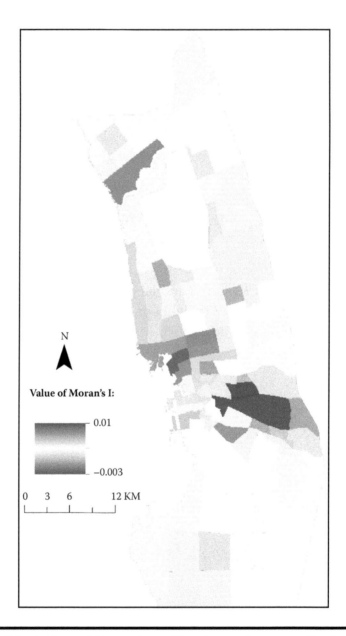

Figure 5.5   Values of Moran *I* resulting from using mosquito data in different Jeddah districts.

values (0.01) of *I* are associated with nonsignificant geographic patterns of spatial clustering, negative values (−0.003) of *I* are associated with a regular pattern, and a value close to zero represents complete spatial randomness (Figure 5.5). Note that areas with different population sizes were given the same weight (0 or 1) in Moran's *I*. The measure yi is the geographic unit's count, which does not include the geographic unit's population information. If a data set has spatial correlations or clustering patterns because of the heterogeneous population sizes, the original Moran's *I* using counts will identify the clustering pattern, which may be caused by the spatial similarity of the population and not the spatial clustering pattern as desired.

5.3 Distribution Analysis

5.3.1 Central Feature

Central feature is a general term used in spatial statistics and distribution analysis. In the analysis of vector-borne diseases such as DF, knowing the most central occurrence of the disease outbreak is key to understanding the source of its spread during subsequent periods of time.

The central feature or the central occurrence of DF is calculated on the premise that, by using the smallest accumulated distance to all other points of occurrence in the study, we can find the most centrally located case. This case can later be considered for special analysis, leading to possible insights into its neighbouring cases.

The type of distance on the basis of which the central feature is to be calculated can be of two types: Euclidean distance and Manhattan distance. Figure 5.6 shows the central features of DF cases as calculated based on the two distance methods. We discuss both of these types of distance in a little more detail.

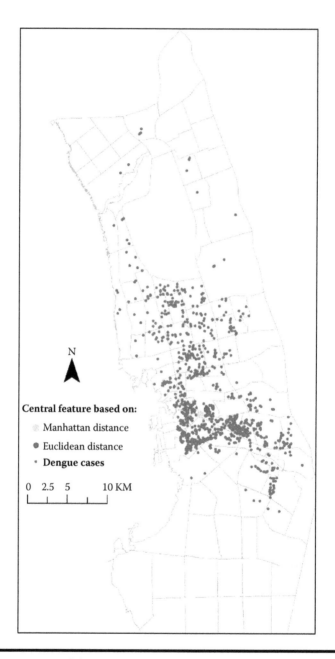

Figure 5.6   Central features of DF case locations based on Euclidian and Manhattan distances.

Euclidean distance is commonly used when the distance to be measured is on a plane. Most scientists and mathematicians may know Euclidean distance as its common name, which is the distance formula.

Mathematically, it is expressed as:

$$dist\left((x,\ y),\ (a,\ b)\right) = \sqrt{\left[(x - a)^2 + (y - b)^2\right]} \quad (5.5)$$

where

$(x,\ y)$ = the x and y coordinates of the source point, respectively

$(a,\ b)$ = the x and y coordinates of the point to be measured to

Euclidean distance, in essence, is the distance that would be measured by a ruler and is derived based on the Pythagorean formula.

A concern when using Euclidean distance is with its accuracy when it is used to measure real distance, such as that between two positions of cases of DF. In this scenario, the distance measured is not on a plane but on Earth's surface with all its associated features. That said, for relatively small distances, the accuracy provided by Euclidean distance is satisfactory (Figure 5.6).

Manhattan distance, also known as the taxicab metric, is the distance measured in Manhattan geometry, which, in contrast to the commonly used geometry or Euclidean geometry, uses absolute differences in the Cartesian coordinates. The name refers to the grid-like layout of most of the streets on the island of Manhattan in New York City.

In simpler terms, the Manhattan distance function computes the distance that would be travelled to get from one data point to the other if a grid-like path is followed. The Manhattan distance between two items is the sum of the differences of their corresponding components.

Mathematically, it can be expressed as

$$dist((x, y), (a, b)) = |x - a| + |y - b| \qquad (5.6)$$

where

(x, y) = the x and y coordinates of the source point, respectively

(a, b) = the x and y coordinates of the point to be measured to

Because of the nature of Manhattan distance, it is predominantly used in metropolitan areas, and given the recent outbreaks of DF in metropolitan areas, its usage can provide relatively more accurate recognition of the central feature in comparison to when Euclidean distance is used (Figure 5.6).

There are many potential uses for the central feature, such as if health officials want to determine a location for a health centre that can best be used to hospitalise people infected by DF. In addition, calculation of the central feature can use a block group feature class, weighted by DF cases, to find the most accessible in the city and make that census block a top candidate. The same method can be implemented for the disease vector.

5.3.2 Standard Deviational Ellipse

Standard ellipses are used to illustrate the dispersion, direction, and orientation of any given distribution. Spatial dispersion can be measured through the observation and calculation of average distances between the features through nearest-neighbour methods, but such calculations only provide a measure of dispersion and do not cater toward a general measure of orientation. A study of mosquito-borne diseases such as DF presents trends by means of features such as mosquito egg counts or cases of DF contraction; here, the standard deviational ellipse can be used to calculate and show the general

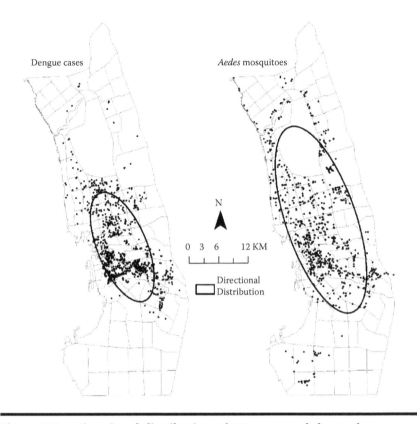

Figure 5.7   Directional distribution of DF cases and the *Aedes* mosquito.

direction of the trend based on these features (Figure 5.7). As shown by Figure 5.7, DF incident distribution and the mosquitoes that transmit the disease have the same directional distribution: to the northwest of the city of Jeddah.

The standard deviational ellipse is based on initially finding the mean aerial centre of the feature set and using it as the origin for axes of the distribution. The standard deviational statistic σ is then calculated.

$$\sigma_y = \sqrt{\frac{\sum_{i=0}^{n}(y_i - \mu)^2}{n}} \tag{5.7}$$

where μ is equal to 0 because it is the origin.

The axes are rotated about the mean, with recalculation for each new position X' of the axes at angle Θ.

$$\sigma_{y'} = \sqrt{\frac{\sum_{i=0}^{n}(y_i' - \mu)^2}{n}}$$

While using an ellipse as a derivative, representation of the distribution has the effect of visually revealing its general location, direction, and dispersion results in guidance toward finding a spatial and temporal correlation in travel and spread routes of the vectors.

5.3.3 Knox Test

There are many available techniques for the spatiotemporal clustering identification of disease locations, but the first technique was developed by Knox (1964). In the Knox test, the separation of pairs of cases is done by less than a user-defined critical space, and time distances are considered and said to be closed in space and time. This will assign a pair of cases to one of four cells in, for example, a 2*2 contingency table that can be shown as far space–far time, near space–near time, near space–far time, and far space–near time. However, the test statistic T_k is computed as the number of pairs of cases that are near to one another in both space and time. The test statistic is compared to the simulated result under the Poisson model, which Knox argues is the sampling distribution of the statistic in the absence of time and space clustering (Pfeiffer et al., 2008). In the literature, this method has been used widely; for example, it was applied to data on cases of childhood leukaemia in northeastern England, identifying significant evidence of space-time clustering (Pfeiffer et al., 2008).

There are two principle limitations of the Knox test. The first is that the choice of critical distances is subjective; second, the critical distance in space does not vary with changing population density. With increasing population density, distance between disease cases would decrease (Jacquez, 1996; Kulldorff and Hjalmars, 1999). Moreover, some problems can be associated with specifying thresholds of proximity in space and time in the Knox test. Therefore, an adaptation was developed. This adaptation does not require unknown critical parameters to be specified. Instead, it allows for a range (from 0 to a maximum space and time differences between any two pairs of cases) to be given for each parameter (Baker, 1996; Kulldorff and Hjalmars, 1999).

With the flexibility in specification of the thresholds with increasing accuracy, the test becomes more powerful. Also, it reduces the Knox test itself when the range for each parameter is reduced to 0 (Pfeiffer et al., 2008).

In addition, the statistical limitations of this method were discussed by Kulldorff and Hjalmars (1999), who proposed a modification of the test. The modification overcomes these problems. The problem was demonstrated using lung cancer cases in New Mexico for the period 1973–1991. At a range of critical distances, an indication after using a standard Knox test showed significant space-time clustering. It has been found that changes in disease risk distribution over time can have a strong impact on the standard test, which is interesting regarding vector-borne diseases, for which cases or vector populations may be culled.

The Knox test is helpful as it can be used by health officers and others involved in disease control to visualise vector-borne diseases, such as the DF spatial and temporal risk patterns found by Sharma et al. (2014). It has been used to provide a better understanding of DF incidence in Trinidad. As a result, decreases in mean distance between DF cases are associated with activity leading to an outbreak, and decreases in temporal distance between cases lead to increased geographic spread of the disease, with an outbreak occurring every 2 years.

5.3.4 Space-Time K Function

Investigating space-time interactions in point process data can be done using existing second-order analysis methods for spatial data. Space-time K functions have properties closely related to or an extension of Knox's statistic (Diggle et al., 1995; Ripley, 1976, 1977; Knox and Bartlett, 1964). K_T is the K function in time; K_S is the K function in space. The K function difference is

$$D(s, t) \text{ is: } D(s, t) = K(s, t) - K_S(s)K_T(t) \tag{5.8}$$

which estimates the cumulative number of disease cases expected within distance s and time interval t of an arbitrarily selected case attributable to the interaction between space and time.

There are many ways to implement this method in a vector-borne disease control system. For example, it can be used to investigate malaria or DF outbreaks in tropical and subtropical areas to reveal strong or weak evidence that disease incidences are clustered, random, or dispersed in both space and time. It can be used to investigate different aspects of, for example, the 2014 DF epidemic and to determine whether it may be attributable either to a contagious process or to other practices, such as human or local climate effects.

5.3.5 Point Buffer

A point buffer is one of the most common spatial analyses. A buffer is a map feature that represents a uniform distance around a disease case or location of a vector that transmits this disease (Figure 5.8). The analyst must consider selection of the feature to buffer from as well as the distance to be buffered when creating that buffer. For example, during an epidemic, restricted space (area) can be declared around an

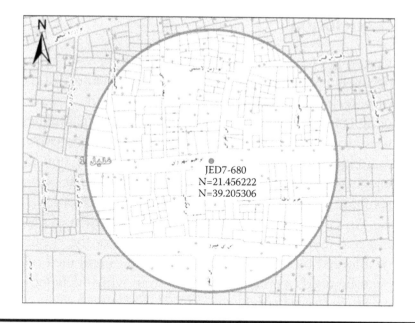

Figure 5.8 **A 500-m buffer zone around a DF case location (large point) to define the spatial parameters of the outbreak area.**

infected house to reduce the risk of spread of a vector-borne disease. The restricted area is structured to establish a buffer of, for example, at least 500 m from an address (longitude and latitude of a house) to protect noninfected areas. The buffer zone is an important area around the infected case home, where transmission is less likely, arising partly because of increases in detection risk related to destroying the favourable environmental conditions for mosquitoes within the home of the infected case and partly because the number of risk opportunities increases with distance from the home.

Making a buffer around an infected DF case can help in allocating adult mosquito- and larvae-favourable areas, which include gardens, water tanks, water pools created through pipelines or drainage system breakage, seepage from slum housing, natural springs, pools and ditches filled with groundwater, and so on.

5.4 Conclusion

The clustering of vector-borne diseases and their vectors can be effectively presented through the use of the variety of analytical techniques discussed. Investigation of the disease clustering is fundamental to epidemiology. These techniques provide considerable information to investigators. The concepts and methods discussed in this chapter provide epidemiologists and health officers with spatial tools that can be useful for surveillance of DF or any other vector-borne disease. An opportunity to specify the health burden of the diseases and their vectors within the hot spots or the clusters can be afforded by the use of these techniques and methods. They can be used to create a platform from which detailed investigations can be conducted on the associated human, environmental, or climatic parameters that are responsible for an increased disease risk or clustering.

References

Baker, R.D. (1996). Testing for space-time clusters of unknown size. *Journal of Applied Statistics,* 23: 543–554.

Bartlett, M.S. (1947). Multivariate analysis. *Journal of the Royal Statistical Society Series B*, 9: 176–197.

Bithell, J.F. (1999). *Disease Mapping Using the Relative Risk Function Estimated from Areal Data. Disease Mapping and Risk Assessment for Public Health*. New York: Wiley; 247–255.

Boots, B., Getis, A. (1988). *Point Pattern Analysis*. Sage University Paper Series on Quantitative Applications in the Social Sciences, series no. 07-001. Thousand Oaks, CA: Sage.

Diggle, P.J., Chetwynd, A.G., Haggkvist, R., Morris, S.E. (1995). Second order analysis of space-time clustering. *Statistical Methods in Medical Research,* 4: 124–136.

Dixon, P.M. (2006). Ripley's *K* function. In: El-Shaarawi, A.-H., Piegorsch, W.W. (Eds.), *Encyclopedia of Environmetrics*. Vol. 3. Chichester, UK: Wiley; 1796–1803.

Earnest A., Chen M.I., Ng D., Sin L.Y. (2005). Using autoregressive integrated moving average (ARIMA) models to predict and monitor the number of beds occupied during a SARS outbreak in a tertiary hospital in Singapore. *BMC Health Service Research*, 5: 36.

Getis, A., Ord, J.K. (1992). The analysis of spatial association by use of distance statistics. *Geographical Analysis*, 24(19): 189–206.

Giesecke, J. (2001). *Modern Infectious Disease Epidemiology*. London: Arnold; 280.

Jacquez, G. (1996). A *K* nearest neighbor test for space-time interaction. *Statistics in Medicine*, 15: 1935–1949.

Khormi, H.M., Kumar, L. (2011a). Examples of using spatial information technologies for mapping and modelling mosquito-borne diseases based on environmental, climatic, socio-economic factors and different spatial statistics, temporal risk indices and spatial analysis: a review. *Journal of Food, Agriculture and Environment*, 9(2): 41–49.

Khormi, H.M., Kumar, L. (2011b). Identifying and vizualising spatial patterns and hotspots of dengue fever: clinically-confirmed cases and female *Aedes aegypti* mosquitoes in Jeddah, Saudi Arabia. *Dengue Bulletin*, 35: 15–34.

Khormi, H., Kumar, L., Elzahrany, R. (2011). Modeling spatio-temporal risk changes in dengue fever incidence in Saudi Arabia: a GIS case study. *Geospatial Health*, 6: 77–84.

Kitron, U., Kazmierczak, J.J. (1997). Spatial analysis of the distribution of Lyme disease. *Wisconsin American Journal of Epidemiology*, 145(6): 558–566.

Knox, G., Bartlett, M.S. (1964). The detection of space-time interactions. *Applied Statistics*, 13, 25–29.

Kulldorff, M., Hjalmars, U. (1999). The Knox method and other tests for space-time interaction. *Biometrics*, 55: 544–552.

Lagrotta, M., Fernandes, T., da Costa Silva, W., and Souza-Santos, R. (2008). Identification of key areas for *Aedes aegypti* control through geoprocessing in Nova Iguaçu, Rio de Janeiro State, Brazil. *Cadernos de Saúde Pública*, 24(1): 70–80.

Lay, J.G., Lin, Z.H., Yap, K.H., Wu, P.C., Su, H.J. (2006). Temperature variability and spatial hotspots of dengue fever occurrence in Taiwan. *Epidemiology*, 17: S485–S485.

Mitchell A. (2005). *The ESRI Guide to GIS Analysis*. ESRI, Press 2.

Nakhapakorn, K., Jirakajohnkool, S. (2006). Temporal and spatial autocorrelation statistics of dengue fever. *Dengue Bulletin,* 30: 177–183.

Pfeiffer, D., Robinson, T., Stevenson, M., Stevens, K., Rogers, D., Clements, A. (2008). *Spatial Analysis in Epidemiology.* Oxford, UK: Oxford University Press.

Ripley, B.D. (1976). The second order analysis of stationary point processes. *Journal of Applied Probability,* 13: 255–266.

Ripley, B.D. (1977). Modelling spatial patterns (with discussion). *Journal of Royal Society Series B* 39: 172–212.

Sharma, K.D., Mahabir, R.S., Curtin, K.M., Sutherland, J.M., Agard, J.B., Chadee, D.D. (2014). Exploratory space-time analysis of dengue incidence in Trinidad: a retrospective study using travel hubs as dispersal points, 1998–2004. *Parasites and Vectors* 7: 341.

Songchitruksa, P., and Zeng, X. (2010). Getis-Ord spatial statistics to identify hot spots by using incident management data. *Transportation Research Record: Journal of the Transportation Research Board,* 2165(1): 42–51.

Tipayamongkholgul, M., and Lisakulruk, S. (2011). Socio-geographical factors in vulnerability to dengue in Thai villages: a spatial regression analysis. *Geospatial Health,* 5(2): 191–198.

Upton, G., Fingleton, B. (1985). *Spatial Data Analysis by Example (Point Pattern and Quantitative Data).* New York: Wiley.

Wen, T.H., Lin, N.H., Lin, C.H., King, C.C., Su, M.D. (2006). Spatial mapping of temporal risk characteristics to improve environmental health risk identification: a case study of a dengue epidemic in Taiwan. *Science of the Total Environment,* 367: 631–640.

Wu, P.C., Lay, J.G., Guo, H.R., Lin, C.Y., Lung, S.C., Su, H.J. (2009). Higher temperature and urbanization affect the spatial patterns of dengue fever transmission in subtropical Taiwan. *Science of the Total Environment,* 407: 2224–2233.

Zeger, S.L., Irizarry, R., Peng, R.D. (2006). On time series analysis of public health and biomedical data. *Annual Review of Public Health,* 27: 57–79.

Chapter 6

Spatial Variation Risk

6.1 Introduction

The creation of graphs allows for better visual feedback during the analysis of any given data set. This in turn permits better understanding of the data itself and highlights intuitively most of the outstanding features. In spatial epidemiology, this is called *disease mapping*. Complicated spatial information and hard-to-detect patterns can be easily visualised through disease maps in ways that could be missed in other representations. Bithell (2000), Diggle (2000), and Lawson (2001) reviewed different perspectives on disease mapping. Spatial epidemiology is focused on three different perspectives (Elliot et al., 2000; Lawson, 2001): (1) disease mapping; (2) disease clustering; and (3) geographical correlation analysis. These do not exhibit cleanly separated boundaries. For instance, a disease map can also be used to report the results of a geographical correlation study or to highlight areas of high or low disease incidence (i.e., locations of clusters in a cluster study) (Berke et al., 2002; Diggle, 2000).

An examination of incidence rates for vector-borne diseases across geographic areas is a valuable public health practice. Investigations into such diseases often include an evaluation of variation in the risk of contraction, which can reveal important details leading to causal explanations. Detailed modelling of the risk of diseases such as dengue fever (DF) serve to assist the authorities responsible for public health provision in taking informed decisions and contribute to the success of prevention programs.

The details presented may refer to DF cases or spatial variations in attribute values such as dengue risk. However, the aim is to produce a geospatial representation of the important effects present in the data while simultaneously removing and filtering noise and irregular values. The resultant maps will show smoothed areas to predict data where none is sampled without introducing significant bias.

Different spatial analytical methods can be used to show the spatial variations in risk. Kernel and point pattern analysis can be used to facilitate visual assessment of the cases or vector presence. Inverse distance weighted (IDW) and kriging methods can be used to describe continuous fields, such as the presence or absence of the *Aedes* mosquitoes that transmit DF.

In this chapter, we show how the kernel and point density methods can be used to examine the level of spatial risk to identify and visualise where values of DF cases or its vector are geographically concentrated and homogeneous. Furthermore, we describe how kriging and IDW techniques can be applied to generate spatially continuous presentations. These types of analyses are particularly helpful in solving problems such as resource allocation. They identify on the basis of extremely pronounced values the risk factor and vectors of these diseases by visualising areas that have been impacted.

6.2 Density Analysis

6.2.1 Point and Kernel

Both the point and kernel density calculate the density of point features around each output raster cell. Generally, a visual analysis of point data density can be performed using a simple display of disease locations, which are used for the smoothing method for generating the disease case density surface. This surface is always represented in a raster format. Kernel density is a method for exploring and displaying spatial patterns in the locations of DF or malaria cases (point data) to reveal areas of high concentrations or to calculate the density of the cases in a neighbourhood around those features.

Possible uses include identifying the density of houses that recorded more than one DF incidence, mosquito reports, or roads or utility lines that can be used for insecticide spray. The population field could be used to weight some disease locations more heavily than others, depending on their context, or to allow one point to represent several incidences. For instance, one address might represent a condominium with six DF cases, or some locations might be weighted more heavily than others in determining overall DF risk levels.

The first method, which is a simple display of disease location, is to be used if the number of recorded cases is limited because it is an easy task to differentiate visually between different densities (Figure 6.1). On the other hand, it should not be used if the number of recorded cases is large (Figure 6.2) as it is not possible to differentiate between areas with high densities. Therefore, it is necessary to apply smoothing methods such as the kernel density.

In the kernel density, a spatial window is moved across the study area, and the density of events (e.g., DF cases) is computed within this window. The size and shape of the window (kernel) can influence the degree of smoothing as well as the mathematical function applied to the values within the window. The window is a circle at a constant radius d from the point (e.g., case location).

The volume under the window corresponds to the population field value for the point or to 1 if none is specified. The density at each output raster cell is calculated by adding the values of all the window surfaces where they overlay the

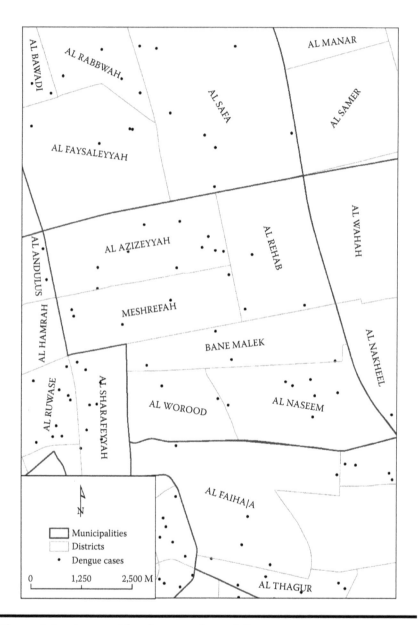

Figure 6.1   Areas with a low number of disease cases.

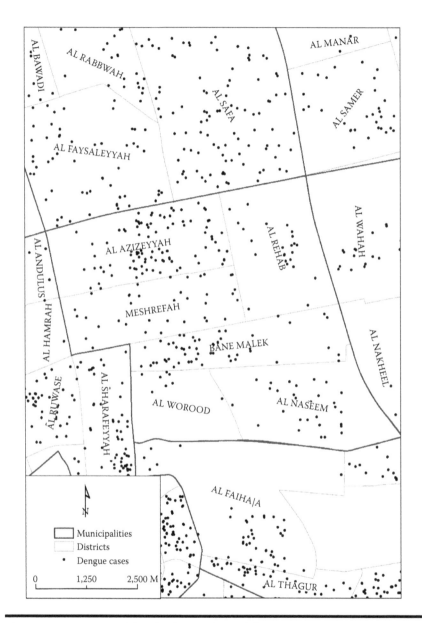

Figure 6.2 Areas with a high number of dengue cases, presenting difficulties in differentiating between areas with high densities.

raster cell centre; in other words, the locations of DF cases within the window are weighted according to their distance from the centre of the window, the points at which the density is being estimated. Cases located near the centre are weighted more highly in comparison to those located geographically far from the centre.

If a population field setting other than 0 is used, each location value (e.g., the number of captured mosquitoes) determines the number of times the point is counted. For example, if the number of captured adult mosquitoes were three, this would cause the point (the location of traps that captured the mosquito) to be counted as three points. The values can be integers or floating points.

The linear unit of the projection definition is based on the selected unit of the input point feature data or as otherwise specified in the output coordinate system environment default setting. The calculated density for the cell is multiplied by the appropriate factor before it is written to the output raster, especially if a unit of area is selected; for example, if the unit of measurement of the input is in meters, the output area's units will default to square kilometres. Comparing a unit scale factor of metres to kilometres will result in the values being different by a multiplier of 4 million (2000 × 2000 m).

An increase in the radius or bandwidth does not have a significant impact on the calculated density values. As a result of increasing bandwidth, more points will fall inside the larger neighbourhood; the number of points will be divided by a larger area when calculating density. The main effect of a larger radius is that the density is calculated considering a larger number of points that can be farther from the raster cell. This results in a more generalised output raster.

To evaluate a realistic implementation of kernel density, using maps of Jeddah's districts, submunicipalities, and the location-based data of adult *Aedes* mosquitoes from the Jeddah municipality, daily mosquito samples were acquired using

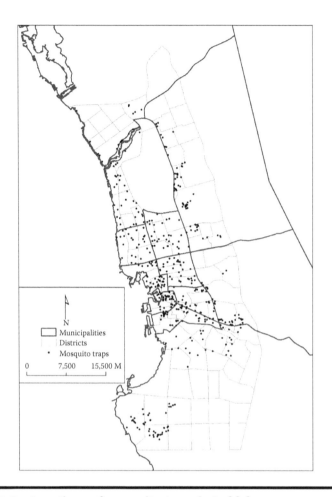

Figure 6.3   Locations of mosquito traps in Jeddah.

black hole traps, which were returned to the mosquito laboratory for filtering and sorting according to species, sex, date of collection, coordinates, and the number of mosquitoes for each location. For the capture of mosquitoes, 504 black hole traps were used. These traps were distributed geographically based on population density and other environmental factors (Figure 6.3). The traps capture mosquitoes by luring them with carbon dioxide.

To visualise the variety of *Aedes* mosquito densities around the study area, kernel density was calculated using 1*1 m as

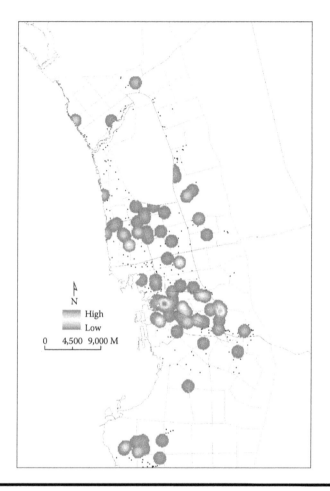

Figure 6.4   Areas with high and low density of mosquitoes.

the cell size and a 10-m search radius. We used the kernel density-mapping technique to create a continuous surface map based on point data. Density surfaces are effective in identifying where vectors are concentrated by highlighting areas of intense activity (Figure 6.4).

In the calculation of the kernel density method, the spatial censoring or edge effects should be taken into account to minimise the bias of estimations close to the study area boundary. In the calculation process, adjusting the area to the overlap of the circular area defined by bandwidth and the study region is introduced to minimise this bias.

6.3 Interpolation

From spatially collected samples of vector-borne disease data, continuous surfaces can be produced to show metrics such as the disease vector risk. These might include, for instance, adult mosquito traps, ground epidemiological surveys that determine DF prevalence or incidences in a sample of communities or districts. An important characteristic of this data type is that its spatial property (location) is not random, as is the case in cluster analysis, but is known and fixed. There are several methods that can be used to convert these points (e.g., DF case locations) into surface information. The two common methods are the IDW method and kriging. The following sections describe how these two methods work and their uses in analysing vector-borne diseases.

6.3.1 *Inverse Distance Weighted Method*

The IDW method calculates missing values from distance-weighted sample measurements to generate a continuous surface that should be a geographically dependent variable. It assumes that the variable being mapped decreases in influence in direct proportion to its distance from its sampled location. As when interpolating a surface of DF for determining risk at sites (Figure 6.5), the DF vector of a more distant location will have less influence because people are more likely to be infected closer to DF vector presence locations.

The inverse of the distance raised to a mathematical power is the main base of the IDW method; the power parameter is important for controlling the significance of known disease locations on the interpolated values based on their distance from the output locations. It is a positive and real number.

More emphasis can be put on the nearest points if the power value is high, and as a result, the greatest influence will come from nearby data. The surface will be less smooth, resulting in sharper detail. When the power increases, the

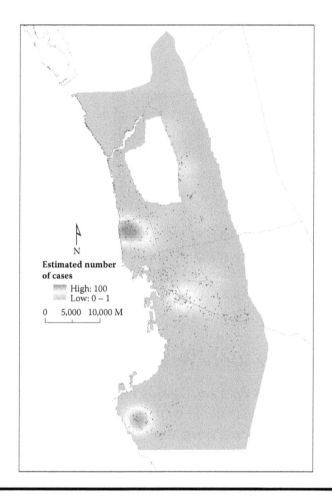

Figure 6.5   Example of using IDW to generate a continuous surface showing disease case density.

values of interpolated locations will approach the value of the nearest sample point, leading to greater influence on points that are farther away, which results in a smoother surface. The IDW formula has no link to any real physical process. As a result, there is no way to determine that a particular power value is too high, and we must consider that, if the distances or the power values are high, the results might be incorrect.

There is no concrete theoretical basis for choosing one power value over others. Selecting a value for the power through concrete calculations can therefore be a complicated process, especially in those studies that have to factor in unknown metrics. Fully understanding the patterns that emerge in the spread of vector-borne diseases will vary according to a large number of additional factors. In the analysis of DF cases, an optimal value for the power can be empirically approximated based on cross validation, which is a common practice to validate the accuracy of an interpolation (Voltz and Webster, 1990). This reduces uncertainty in the results. This can be done by arbitrarily selecting a subset of the points at which the number of, for example, mosquitoes identified is known and using them to calculate an approximate value for points whose values are already known using IDW interpolation. An estimated value of the power can be chosen by varying it to find the best match between the calculated values and the known number of cases at points (Figure 6.5).

This matching can be performed using the root mean square prediction error (RMSPE) (Mevik and Cederkvist, 2004). RMSPE measures the expected root of the squared distance between the value obtained through interpolation and the actual value. Exhaustive cross validation can also be performed, such that the value of each known point is interpolated and matched with its known value, leading to better understanding of the accuracy of the parameters used with the interpolation model.

Controlling the characteristics of the interpolated surface is important for improving the result, which can be accomplished by limiting the input points (e.g., the locations of the *Aedes aegypti* mosquito) that are used in the calculation of each output cell value. Some locations or points can be eliminated, especially when they are far away from the cell location, because they may have poor or no spatial correlation. It is important to limit the disease case points, which will be used in the calculation of each output cell value, to control the characteristics of the interpolated surface. In other words, the question arises regarding whether spatial closeness is comparable to similarities in value. This is crucial because, if the number of DF cases is spatially correlated, then these cannot be calculated as independent observations. If a correlation in the observations is unaccounted for, then the predictions and risk analysis of such missing data will be modelled in an erroneous manner. Moreover, it is sometimes necessary to specify the number of points or locations to use directly or to specify a fixed radius within which points will be included in the interpolation.

The number of disease locations (points) should be specified to compute the value of the interpolated area, which can make the radius distance vary for each interpolated area. This also depends on the distance needed to investigate each interpolated area to achieve the required amount of input data. Based on the density of the measured points close to the interpolated area, some neighbourhoods will be either small or large. It is also important to specify a maximum distance to predict the value for that area depending on the measured disease points within the determined maximum distance. If the disease prevalence contains a significant degree of variation, it is necessary to select a small neighbourhood and a minimum number of points to preserve a constant radius distance for each interpolated area. Should the measured DF points or locations not be spread equally, using a specified fixed search radius will be the optimum methodology because the likelihood of the existence of a different number of measured points used in the different neighbourhoods for the various predictions is highly probable (Figure 6.5).

Mathematically, the estimated values can be calculated using the following equation:

$$\mu_0 = \frac{\sum_{i=1}^{N(\mu_0)} \dfrac{1}{d_i^p}}{\sum_{i=1}^{N\mu_0} d_i^p} \tag{6.1}$$

where

μ_0 = the estimated value at (x_0, y_0)
μ_i = a neighbouring data value at (x_i, y_i)
d_i = the distance between (x_0, y_0) and (x_i, y_i)
p = the power
$N(\mu_0)$ = the number of neighbouring data points in the vicinity of μ_0

6.3.2 *Kriging*

Kriging is a method of statistical spatial interpolation widely known for spatial prediction purposes. It evaluates the distances to the control points and the measurement of spatial autocorrelation between the control points. Originally, it was used to describe continuous-scale outcome variables. For example, it can be used to find the specific concentration of mosquito eggs in the soil. Various implementations of kriging have been proposed in the past.

In kriging, the assumption reflects the way in which the direction and distance between sample disease vector points are spatially correlated and can be used to explain variations in the surface. A mathematical function to a specified number of points, or all points within a specified radius, is fitted to determine the value for each location. When the spatially correlated distance or directional bias in the data is known, kriging will be an appropriate tool.

There are some advantages to using kriging, such as compensating for the effects of diseases or their vector clusters. For example, assigning individual disease points within a cluster is weighted less than isolated data points (or treating clusters more like single points), giving an approximation of the estimation error (kriging variance). The availability of an estimation error provides a basis for stochastic simulation of possible realisations. In simple terms, kriging is often used to find a set of weights for estimating the variable value at the location that does not show *A. aegypti* present from values at a set of neighbouring recorded locations. Increasing the distance between recorded locations leads to a decrease in the weight on each data point.

To interpolate risk surface for diseases or vectors using kriging, a number of steps must be considered, such as an exploratory statistical analysis of the data, variogram modelling, creating the surface, and (optionally) exploring a variance surface.

The general kriging mathematical formula can be stated as follows:

$$\hat{Z}(s_0) = \sum_{i=0}^{n} \lambda_i Z(s_i) \tag{6.2}$$

where

$Z(s_i)$ = the measured value at the *i*th location

λ_i = an unknown weight for the measured value at the *i*th location

s_0 = the prediction location

N = the number of measured values

The weights are calculated using their general spatial arrangement besides the distance between points and the location of the prediction. Knowing the spatial autocorrelation is a prerequisite for utilizing spatial arrangement in the calculation of the respective weights.

Two major steps are involved in kriging: (1) understanding the dependency rules and (2) finally making the predictions. Dependency rules are inferred using spatial autocorrelation, which is calculated by variograms and covariance functions. These are dependent on the underlying model of correlation, which is discovered by trying different models. This step is called *model fitting*.

Variography is used for model fitting. First, a graph of the empirical variogram is determined using the following equation:

$$\text{Semivariogram}(\text{distance}_h) =$$
$$0.5 * \text{average}\left(\left(\text{value}_i - \text{value}_j\right)2\right) \tag{6.3}$$

where the pairs of locations, separated by distance *h*, are made and used as parameters. Commonly, the distances between these numerous pairs of locations are unique.

Fitting a model based on the previously formed empirical semivariogram utilizing the points is a core step in spatial description and spatial prediction. Prediction of values at locations where samples were not taken (i.e., interpolation) is the primary use of kriging. Information regarding the spatial autocorrelation is gleaned from the empirical semivariogram; information with respect to different parameters, such as direction and distance, and all of their possibilities is not as yet thereby gained. For the resolution of this unknown value, fitting a model is the key that helps to ensure that kriging-based predictions are always situated with nonnegative variances (i.e., a continuous function or curve of the empirical semivariogram). At a certain level, model fitting is comparable to regression analysis in which a function creating a curve or line is matched with the known points.

The first step to fitting a model is to select a function that matches our model (e.g., a spherical model that increases and flattens at a certain range, i.e., sufficiently large distances). There will almost always be points that, instead of fully matching, deviate from the empirical semivariogram, with some points above the curve of the model and some of them possibly below the curve. Generally, however, if we add the distance of the points above the curve and compare them to the distance of the points below the curve, they will be evenly

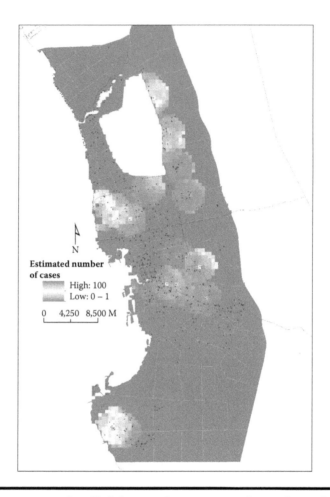

Figure 6.6   Example of kriging used to generate the surface of disease density using spherical semivariogram model fitting.

matched. There are many semivariogram models to choose, with choice based on the data.

Commonly used functions for semivariogram model fitting include spherical (Figure 6.6), circular (Figure 6.7), exponential, Gaussian, and linear. Unknown values are interpolated on the basis of the selected model and are significantly affected by it, especially when there are stark differences in the curve at the origin. The influence of the neighbouring points increases with a steeper curve at the origin, which results in a more aberrant

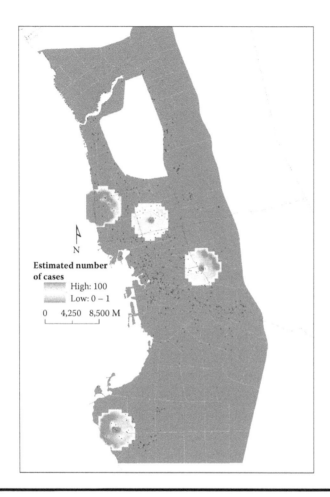

Figure 6.7   Example of kriging used to generate the surface of disease density using circle semivariogram model fitting.

surface being generated. Each model fits the scenarios of different diseases more accurately based on its characteristics.

It is a general geospatial trait that closer locations are likely to be similar, so that points that are closer will have a much smaller difference with respect to the number of DF cases compared to those that are far apart. After the pairs of locations are plotted onto a graph, a model is fitted based on the given characteristics. Range, sill, and nugget are commonly used to describe these models.

On observing the model of a variogram, it is noticeable that the model curve flattens after a certain distance is covered. The range is the distance where the curve flattens. Thus, locations sampled for DF cases that are within the range are to be autocorrelated; the locations that are farther apart than the range are not.

The sill, on the other hand, is said to be the value at which the range is obtained. In addition, a partial sill is the difference between the sill and the nugget.

Conceptually, when lag is 0, that is, when the distance that separates the points is zero, the semivariogram value is also 0. In practice, however, the semivariogram displays a nugget effect (i.e., a value greater than zero) when the separation distance is made infinitely small. The nugget's value is the value on the y axis at which the model curve intercepts it.

The effect of having a nugget can be caused by two possible factors. One is a simple error in measurement. The other can be spatial variation in sources at a distance less than the interval between sampling. Measurement error can arise for various reasons (e.g., malfunctioning of the traps or miscalculation of disease cases). Thus, any variation smaller than the sampling interval will reveal itself in the form of a nugget. Therefore, before collection of data on disease cases or setting traps for the vector, comprehending the disease's characteristics is of utmost importance to produce reliable results.

Having discovered the dependence and autocorrelation of our data and finishing its first use (i.e., calculating distance and modelling the spatial autocorrelation by means of the spatial information), predictions can then be made based on the utilisation of the fitted model, at which point the empirical semivariogram no longer remains in use.

Based on the sampled data, predictions can now be made by interpolation. Similar to interpolation as observed in the IDW technique, kriging also utilises calculated weights based on the nearer sampled values to predict unsampled locations.

Values obtained through kriging are also observed to be influenced more by geographically closer sampling locations, similar to the IDW method. In hindsight, the weights based on kriging are felt to be more sophisticated in comparison to their counterparts derived using the IDW technique. The weight calculation algorithm used in that method is relatively simple, whereas a semivariogram is the source of kriging-based weights; it was developed by looking at the spatial characteristics of the disease as observed in raw data. In the creation of a continuous surface for the disease, the interpolated values are derived for each location or health care unit in the observed area based on the semivariogram and the geographical arrangement of the sampled cases and vector population.

6.3.3 Other Common Trend Surface Functions

Trend surface analysis (TSA), also referred to as trend surface mapping, is another interpolation method that has been used in the past for spatial analysis of the propagation of such diseases as plague (Maidana and Yang, 2009) and rabies by raccoons and other wildlife (Lucey et al., 2002; Moore and Carpenter, 1999). TSA is used to study diffusion processes in space and time; in our context, this will apply to the diffusion of the vector or its eggs. It is a least-squares regression method. By mapping the specific timing of events at each point (longitude, latitude), a surface pattern can then be created. The TSA procedure details have been described elsewhere (Unwin, 1975). In general, this method has been used in conjunction with a model with power series polynomials, fitting linear, quadratic, cubic, and higher-order trend-surfaces to the data (Moore and Carpenter, 1999). The order of the polynomial chosen as the model determines the characteristics of the trend surface, such as shape and flexibility (Lucey et al., 2002). Trends are limited to a plane through the data points when a first-order polynomial is used. When a curvature over the entire data set is required, a second-order polynomial is used. To allow a much more local curvature to the surface, higher-order polynomials are favoured (Moore, 1999).

In the study of vector-borne diseases such as DF, a polynomial surface must be fitted according to the set of spatially distributed times that clinical cases will have detected across the area under surveillance with respect to the municipalities (Equation 6.4). While conducting a study with TSA, the area where the disease of origin is identified as is the area of first infection (AFI) and its date. After obtaining the AFI for DF and its date of introduction, the model type has to be decided.

As an example to estimate linear and quadratic surfaces by least squares, we take the following model into consideration:

$$t = \beta_0 + \beta_1 X + \beta_2 Y + \beta_3 X^2 + \beta_4 XY + \beta_5 Y^2 + \varepsilon \qquad (6.4)$$

where

t = the number of days to disease introduction
βi = the fitted parameters
X and Y = the geographical parameters of the municipality centroids
ε = the error term representation

The biggest limitation when significance testing for TSA is the omnipresence of spatial auto correlation, which, as discussed in previous sections, is an inherent property of spatial data (Unwin, 1975).

Neighbourhood matrices can utilise autoregressive models to incorporate spatial autocorrelation, which represents one possible solution to deal with spatial autocorrelation. At each location i and those at neighbouring location j, these matrices can map the relationship between the response values if conditional autoregressive (CAR) models are used or residual if simultaneous autoregressive (SAR) models are used. On the basis of where the point at which the spatial autoregressive process is believed to occur, three different forms can be taken for the SAR models (Dormann, 2007). It is assumed in the spatial error model (SAR$_{err}$) that the autoregressive process occurs not in the response or predictor variables but only in the error terms. In the lagged response model (SAR$_{lag}$), the autoregressive process is assumed to occur in the response variable, whereas both response and predictor variables are assumed to be affected by spatial autocorrelation. Along with ordinary least squares (OLS) regression, these three SAR models (SAR$_{err}$, SAR$_{lag}$, and SAR$_{mix}$) were tested for performance (Kissling and Gudrun, 2007). It has been observed that SAR$_{err}$ models performed most reliably of all the SAR models and generally produced favourable results in various other cases (i.e., regardless of the type of spatial autocorrelation and the models that were chosen by SAR$_{min}$, R^2 [the coefficient of determination of a linear regression], or AIC [Aiaike information criterion]). On the other hand, unexpected biases in parameter estimates and very weak type I error control were exhibited by OLS, SAR$_{lag}$, and SAR$_{mix}$ models. In conclusion, the SAR$_{err}$ model was found the most favourable to factor in the spatial autocorrelation found in the model residuals.

6.3.4 Radial Basis Function

Often, data points need to be interpolated from data that does not have very erratic values. Such data also requires that the interpolated continuous surface be smooth. Radial basis function (RBF) methods are a set of exact interpolation techniques; that is, the generated continuous surface passes through all of the known data points provided to it as data input.

There are five common RBF methods:

- Multiquadratic function
- Thin-plate spline
- Inverse multiquadratic function
- Completely regularised spline
- Spline with tension

The RBF method is not suitable when the horizontal variance of the data points is great or when the reliability of the values of the sample data has been brought into question and has a significantly high probability of error and uncertainty.

The smooth nature of the calculated continuous surface using RBF methods make them suitable for lightly varying data, such as weather conditions data. In the following section, we discuss the rationale behind using weather data for dealing with vector-borne diseases such as DF and their vector (i.e., *Aedes*), with particular focus on the general spline function.

6.3.5 Spline

Besides directly sampling the number of cases of vector-borne diseases such as DF, it has been noted that global climate change has significant implications for and a potentially serious impact on the dynamics of the spread of such diseases in spatial and temporal dimensions. Factors such as competition, predation, parasitism, and other density-dependent factors have a visible effect on the presence of vectors such as *Aedes*. Large-scale density-independent factors also have a significant, often recurring, impact. For example, the type of habitat of the vector species may be identified in different weather conditions (Holdridge, 1971), as can the suitability of arthropod vectors in such environments (Rogers et al., 1996; Duchateau et al., 1997; Robinson et al., 1997; Cumming, 2002).

It has been predicted that the current warming phase of Earth will have a significant impact on the redistribution of many vector-borne diseases. This is because the existence of vectors and pathogens is highly connected to climate change (Reeves et al., 1994; Patz et al., 1996).

In general, climate change therefore has a significant impact on disease vectors, but specifically, warming of the general climate has been predicted and is known to increase the intensity of transmission to exacerbate the spread of mosquito-borne diseases such as malaria and DF (Martens et al., 1995; Tanser et al., 2003; Martens, 1998; Hales et al., 2002; Lindsay and Martens, 1998). To be more precise, regions that were previously thought to be uninhabitable by arthropod vectors such as *Aedes* may become habitable with climate change besides shortening the pathogen incubation period and increasing the occurrence of bites and rate of reproduction (Shope, 1991; Patz et al., 1996).

Thus, to study and create a surface for the geographic weather data, spline can be used as a method for interpolation. Spline interpolation estimates values using a mathematical function that minimises overall surface curvature. The resultant surface is smooth enough to allow it to pass through all the sampled data points.

Conceptually, we can consider a sheet of stretched rubber that spline bends such that it passes though the data points while maintaining minimum cumulative curvature of the surface. Thus, an underlying mathematical function is specified in these constraints. Spline interpolation is commonly used for gently varying surfaces, such as weather data. Therefore, formally the two conditions for spline interpolation are that the surface must pass exactly through the known data and the curvature of the surface must be minimal; that is, the cumulative value of the squares of the second derivative of the surface at each must be as low as possible.

Points calculated using spline interpolation can be expressed as the following equation:

$$S(x, y) = T(x, y) + \sum_{j=i}^{n} \lambda_j R(r_j) \qquad (6.5)$$

where

j = 1, 2, ..., n
n = the number of points
λ_j = coefficients found by the solution of a system of linear equations
r_j = the distance from the point (x, y) to the jth point
$T(x,y)$ and $R(r_j)$ = differently defined items based on the selected point

References

Bithell, J.F. (2000). A classification of disease mapping methods. *Statistics in Medicine,* 19(17–18): 2203–2215.

Berke, O., von Keyserlingk, M., Broll, S., Kreienbrock. L. 2002. Zum Vorkommen von Echinococcus multilocularis bei Rotfüchsen in Niedersachsen: Identifikation eines Hochrisikogebietes mit Methoden der räumlichen epidemiologischen Clusteranalyse. *Berl Munch Tierarztl Wochenschr.* 115:428–434.

Cumming, G.S. (2002). Comparing climate and vegetation as limiting factors for species ranges of African ticks. *Ecology,* 82(1): 255–268.

Diggle, P.J. (2000). Overview of statistical methods for disease mapping and its relationship to cluster detection. In: Elliott, P. et al. (Eds.), *Spatial Epidemiology: Methods and Applications,* Oxford, UK: Oxford University Press,, 87–103.

Dormann, C.F. (2007). Effects of incorporating spatial autocorrelation into the analysis of species distribution data. *Global Ecology and Biogeography,* 16, 129–138.

Duchateau, L., Kruska, R.L., Perry, R.D. (1997). Reducing a spatial database to its effective dimensionality for logistic-regression analysis of incidence of livestock disease. *Preventive Veterinary Medicine,* 32(34): 207–218.

Elliot, P., Wakefield, J.C., Best, N.G., Briggs, D.J. (Eds) (2000). *Spatial Epidemiology: Methods and Applications.* Oxford, UK: Oxford University Press.

Hales, S., de Wet, N., Maindonald, J., Woodward, A. (2002). Potential effect of population and climate changes on global distribution of dengue fever: an empirical model. *Lancet,* 360(9336): 830–834.

Holdridge, L.R. (1971). *Forest Environments in Tropical Life Zones: A Pilot Study.* New York: Pergamon Press.

Kissling, W.D., Gudrun, C. (2007). Spatial autocorrelation and the selection of simultaneous autoregressive models. *Global Ecology and Biogeography.* doi:10.1111/j.1466-8238.2007.00334.x

Lawson, A.B. (2001) Tutorial in biostatistics: disease map reconstruction. *Statistics in Medicine,* 20: 2183–2203.

Lindsay, S.W., Martens, P. (1998). Malaria in the African highlands: past, present and future. *Bulletin of the World Health Organisation,* 76: 33–45.

Lucey, B.T., Russell, C.A., Smith, D., Wilson, M.L., Long, A., Waller, L.A., et al. (2002). Spatiotemporal analysis of epizootic raccoon rabies, propagation in Connecticut, 1991–1995. *Vector Borne Zoonotic Diseases,* 2: 77–86.

Maidana, N.A., Yang, H.M. (2009). Spatial spreading of West Nile Virus described by traveling waves. *Journal of Theoretical Biology,* 258: 403–417. doi:10.1016/j.jtbi.2008.12.032

Martens, P. (1998). *Health and Climate Change: Modelling the Impacts of Global Warming and Ozone Depletion.* London: Earthscan.

Martens, W.J., Niessen, L.W., Rottmans, J., Jetten, T.H., McMichael, A.J. (1995). Potential impact of global climate change on malaria risk. *Environmental Health Perspectives,* 103(5): 458–464.

Mevik, B.-H., and Cederkvist, H.R. (2004). Mean squared error of prediction (MSEP) estimates for principal component regression (PCR) and partial least squares regression (PLSR). *Journal of Chemometrics,* 18(9): 422–429.

Moore, D.A. (1999). Spatial diffusion of raccoon rabies in Pennsylvania, USA. *Preventive Veterinary Medicine,* 40: 19–32. doi:10.1016/S0167-5877(99)00005-7

Moore, D.A., Carpenter, T.E. (1999). Spatial analytical methods and geographic information systems: use in health research and epidemiology. *Epidemiology Review,* 21: 143–161.

Patz, J.A., Epstein, P.R., et al. (1996). Global climate change and emerging infectious diseases. *JAMA,* 275(3): 217–223.

Reeves, W.C., Hardy, J.L., et al. (1994). Potential effect of global warming on mosquito-borne arboviruses. *Journal of Medical Entomology,* 31(3): 323–332.

Robinson, T., Rogers, D., et al. (1997). Mapping tsetse habitat suitability in the common fly belt of southern Africa using multivariate analysis of climate and remotely sensed vegetation data. *Medical and Veterinary Entomology,* 11(3): 235–245.

Rogers, D.J., Hay, S.I., et al. (1996). Predicting the distribution of tsetse flies in West Africa using temporal Fourier processed meteorological satellite data. *Annals of Tropical Medicine and Parasitology,* 90(3): 225–241.

Shope, R. (1991). Global climate change and infectious diseases. *Environmental Health Perspectives,* 96: 171–174.

Tanser, F.C., Sharp, B., et al. (2003). Potential effect of climate change on malaria transmission in Africa. *Lancet*, 362(9398): 1792–1798.

Unwin, D.J. (1975). *An Introduction to Trend Surface Analysis.* Concepts and Techniques in Modern Geography. Norwich, UK: Geo Abstracts.

Voltz, M., Webster, R. (1990). A comparison of kriging, cubic splines and classification for predicting soil properties from sample information. *Journal of Soil Science,* 41(3): 473–490.

Chapter 7

Modelling Associations between Mosquito-Borne Diseases and Environmental and Socioeconomic Factors

7.1 Introduction

Health and related disease authorities must take into account spatial modelling aspects to understand a disease and its vector occurrences, as well as its associations with environmental and socioeconomic factors. In tropical and subtropical areas, mosquito-borne diseases are a major health issue. No medical vaccinations exist for some of these diseases, such as dengue fever (DF); therefore, the best way to avoid their impact is to model their spatial associations with the environment to help monitor their vectors and their environmental conditions. Geographic information systems (GISs) support the modelling of this spatial association using entomological, epidemiological, environmental, and socioeconomic data, as well as the production of risk models for vectors and diseases. Because of their abilities to link different types of information on environmental, climatic, and socioeconomic factors for a given area, they can be used in different spatial statistics analyses and for the development of spatial databases that can be applied to a wide range of public health programs.

Knowledge of the type of outcome model variables is important because it helps determine the model techniques to be used as well as the options available to account for spatial dependence. This chapter describes the use of GIS modelling approaches in the field of mosquito-borne diseases. Modifications of modelling solutions for use in disease-endemic environments must be a part of the new focus of mosquito-borne disease research. Furthermore, this chapter is divided into two main sections. The first outlines the disease, environmental, and socioeconomic data; in the second section, we discuss how to model the association using geographically weighted regression (GWR), ordinary least squares (OLS), Poisson regression, linear regression, and multiple regressions.

The objective is to provide ideas regarding the available methodologies and to highlight the additional information that might be used to understand further the determinants of mosquito-borne diseases in human populations.

7.2 Verifying Required Data

7.2.1 Disease Data

To provide disease data, disease surveillance is required to identify the spread of disease and to monitor it to establish patterns of progression. The main aim of disease surveillance is to predict, observe, and minimize the harm caused by outbreak, epidemic, and pandemic situations, as well as to increase knowledge regarding which factors contribute to such circumstances. Therefore, the practice of disease case reporting is the main aspect of modern disease surveillance. Methods of reporting and registering cases of diseases have transitioned from manual record keeping to instant worldwide Internet communication. Disease case numbers could be collected from hospitals that would see most of the occurrences, collated, and eventually made public.

In modern times, reporting and registering cases have changed dramatically because of new devices and technologies, and some local health authorities in Saudi Arabia and other countries routinely report cases of mosquito-borne diseases within different periods, ranging from days to hours after the incidence. These disease case records, such as for DF, are an important source of spatiotemporal data for public health. These records include age, sex, nationality, district, geographic coordinates, social characteristics, and time of disease onset for each case. Information about the disease case and the infection processes are often accurate, but data about the historical movements of the case before infection by the disease can have errors and inconsistencies.

The disease records often have the patient address or *x, y* coordinates, a spatial identifier for GIS modelling. This information can be used to model environmental and socio-economic influences on disease distribution, such as the clustering of cases in relation to swamp sites. As well, health authorities collect and report data on deaths using mortality records, including demographic characteristics, information about the immediate cause of death, and contributing factors (Friis and Sellers, 2009; Cromley and McLafferty, 2012). The death information also includes two address types based on geographical context, such as the place of death and the usual address of the deceased.

Addresses dependent on statistics information present some challenges to the GIS modellers or researchers, including coding the address incorrectly, which make it impossible to recognize the spatial locations. In addition, many health and related disease authorities do not provide the address or geographical location of a patient because of privacy and confidentiality reasons. They only provide aggregate data, such as a zip code, district, or subdistrict, which is useless for analysing or modelling point locations. The last challenge is that spatial coordinates do not accurately represent the environment of the patient before and during the infection. Neither the spatial address nor the coordinates are correct because the relevant exposure may have occurred in a place other than the residence. For disease data, it is problematic given that the conditions that lead to infections can result from lifelong exposures and behaviours.

Disease surveillance involves monitoring disease distribution data gathered for a specific population and geographical area (e.g., gathering DF cases in Jeddah, Saudi Arabia). The different types of disease data that can be gathered range from information surveys and research projects to information gathered by government agencies and health care providers. The following paragraphs explain reportable disease data from the US National Notifiable Diseases Surveillance System (NNDSS) and disease registries from other countries (Friss and Sellers, 2009; Cromley and McLafferty, 2012).

Information provided on diseases and certain reportable health conditions is called *reportable disease data*. Mosquito-borne diseases are important health issues and the focus of health surveillance in many tropical and subtropical countries. In Saudi Arabia, for example, the authority to require notification of disease cases resides in the province or county health affairs departments. Physicians are required to report cases of specified notifiable diseases to province or local health departments. At the local level, the two available data sources are notifiable disease reports and vital records.

In Saudi Arabia, mosquito-borne disease surveillance systems are usually operated by the Centres for Borne-Disease Vectors Control (e.g., the National Centre for Borne-Disease Vectors Control in Jizan) in collaboration with epidemiologists and municipalities. These centres also report nationally and internationally in compliance with the World Health Organisation (WHO) International Health Regulations. The main mosquito-borne diseases are DF, malaria, and Rift Valley fever (RVF), which are designated as notifiable at the county and province levels. The list of county- and province-reportable mosquito-borne diseases and other conditions changes periodically, and reporting practices may differ from province to province.

Two common surveillance systems are active and passive. The active system gathers data by searching and maintaining periodic contact with a provider; the passive system depends on reports given by providers. Both have implications regarding completeness of the data, especially the spatial database quality. A mechanism for completing and correcting spatial or descriptive information from a reportable disease can be offered by the active surveillance system, which includes address data used as a geographical identifier. To ensure privacy and confidentiality, the health affairs department in Saudi Arabia releases reportable disease statistics at the county level. Some local health affairs departments provide information for a certain geographical area or coordinates for controlling purposes as long as the beneficiary agrees to maintain privacy and confidentiality.

However, for the collection of information on specific diseases, disease registries are always centralized. For example, health and local authorities manage the cancer registries. There are similarities between disease registries and reportable disease data when using the reporting system, with health providers reporting occurrences to the appropriate province or local registry. Some registries collect reports; others seek case information. In addition, keeping longitudinal information is the aim of some registries so they can follow patients after diagnosis, which is important for tracking changes in health status and treatment regimens.

Reportable disease data, the NNDSS, and disease registries are sources of disease data that are necessary for modelling spatial associations between mosquito-borne diseases and environmental and socioeconomic factors, especially to identify disease risk spatial patterns. Implementing GIS surveillance systems at the province or county level increases the likelihood that a disease case database will include cases identified using different case definitions and surveillance methods.

7.2.2 Environmental Data

Vectors of diseases, such as mosquitoes, cause illness and death in more than 300 million clinical cases every year around the world. Many vector-borne diseases, including malaria, DF, and RVF, are transmitted by different mosquito types, creating great public concern (Troyo, 2007; Khormi and Kumar, 2011a). The environmental data related to mosquito breeding sites can be verified or can be extracted using satellite images; depending on the biological criteria and environmental conditions, the mosquito density can be determined. Remote sensors can provide imagery with low or high spatial resolution that can be utilized to model these conditions. The result of these models provides information on the spatial and temporal characteristics of preferred mosquito environmental settings. The potential links between remotely sensed environmental conditions and mosquito habitats are summarized in Table 7.1 (Beck et al., 2000; Khormi and Kumar, 2011a).

Remote-sensing imagery, or a satellite image, is a viable predictor of mosquito-borne disease transmission. For example, it has been used to determine that higher malaria incidence rates were associated with broadleaf hill forests, agricultural land, wetland, and vegetation types (Hakre et al., 2004). It can show how a ground mosquito survey shows that malaria-carrying mosquito populations are prevalent in rice paddies and forestlands (Sithiprasasna et al., 2005; Rakotomanana et al., 2010). In addition, high-resolution images, such as SPOT images, can be used to show the

**Table 7.1   Remotely Sensed Environmental Factors
and Mosquito Habitats**

Remotely Sensed Factors	Description	Remotely Sensed Factors	Description
Deforestation	Habitat creation (sunlight pools)	Vegetation/crop style	Breeding, resting, and feeding habitats
Flooded forests	Mosquito habitats	Wetland	Mosquito habitats
Flooding	Mosquito habitats	Soil moisture	Breeding habitat
Permanent water	Mosquito habitats	Canals	Dry season mosquito habitat
Vegetation green up (response to a rainfall events)	Timing of habitat creation	Human settlements	Source of infected humans, population at risk for transmission

Source: Beck, L.R., Lobitz, B.M., Wood, B.L. (2000). *Emerging Infectious Diseases*, 6(3): 217–227.

probability of the presence of malaria- or DF-carrying mosquitoes depending on the amount of forest between buildings and waterways, the distance of the buildings from the waterways, and the altitudes above specified waterways. The result can verify the closest areas to the water in both distance and altitude that have the highest probability of mosquito presence (Khormi and Kumar, 2011a).

Climatic variables can be considered another dimension of the environmental factors. Temperature, relative humidity, and precipitation play important roles in the mosquito population density as well as in the replication and transmission of diseases. It has been found that mosquitoes are critically dependent on climate for their survival and development. In general, temperature ranges from 14°C to 18°C at the lower end and 35°C to 40°C at the higher end can lead to greater transmission occurrences, especially for DF (Githeko et al., 2000). Development increases in warmer temperatures, raising the chances of disease transmission; the reproduction rates and replications of diseases are slower in cooler temperatures (Githeko et al., 2000; Dye and Reiter, 2000; Monath and Tsai, 1987; Epstein, 2000).

In general, high amounts of precipitation lead to increases in the number of breeding sites, and humidity is often overlooked as a factor in the life cycle of mosquitoes and in disease replication and transmission. Relative humidity is increased by rainfall, particularly following drought. Relative humidity strongly affects flight and the subsequent host-seeking behaviours of mosquitoes (Day and Curtis, 1989; Khormi and Kumar, 2011a; Takken and Knols, 2007).

Many studies have used the climate variable impact for mosquito-borne disease distributions. For instance, Moore (1985) predicted an *Aedes aegypti* abundance from climatological data, finding that temperature was not a good indicator of larval abundance. In this study, the amount of rainfall and the number of rainy days were useful predictors of larval abundance. In Mexico, the average temperature during the rainy season related strongly to the estimated risk of DF infection (Koopman et al., 1991). In addition, there was a significant relationship between humidity and infection.

The weekly DF morbidity and monthly rainfall data of Trinidad were analysed, and the results suggested there is a significant relationship between the temperature and the DF incidence rates and a slightly negative correlation with the rainfall rate (Wegbreit, 1997). Lindsay and Birley (1996) developed a model to illustrate the impact of small increases in the temperature on the transmission of *Plasmodium vivax* malaria. The result showed that, at low temperatures, small increases lead to large reductions in the time for malaria development and hence a disproportionate increase in transmission. They considered different aspects of the influence of global environmental change and emphasized that the direct effects of temperature increase on malaria would be most obvious in the highland areas of Ethiopia, Madagascar, and Kenya (Molyneux, 1997).

7.2.3 Socioeconomic Data

To build good control and prevention systems, socioeconomic factors must be understood. There is a known impact of socio-economic factors that affect the prevalence of diseases. For example, in some countries, such as Saudi Arabia and Brazil, DF affected mainly adults, but an increase in occurrences was occasionally observed in younger age groups. The spatial relationship between disease cases and population density can be analysed using GISs to find a strong positive or negative spatial relationship between DF prevalence and population density (Khormi and Kumar, 2011b).

Different socioeconomic factors can be verified and characterized to model the risk of diseases. In this chapter, we provide an overview of the socioeconomic factors that can be collected easily from census departments, such as population numbers, gender, and nationality, or from using satellite images, including population density and neighbourhood quality.

In modelling the relationship between disease cases and the annual population in each zone, GIS layers must be created and projected to reference the data of a study area or zone. The population and population breakdown (age, sex, nationality) of each zone, for example, and the number of cases can then be attributed to that layer.

By extracting the inhabited area or zone from satellite images and using population numbers, the population density of each zone can be characterized or calculated. For example, IKONOS images for 2014 (at 1-m spatial resolution) of any zone can be used to digitize the inhabited areas in each zone to assist in deriving a proxy indicator for an accurate population density value (Figure 7.1). The population number can

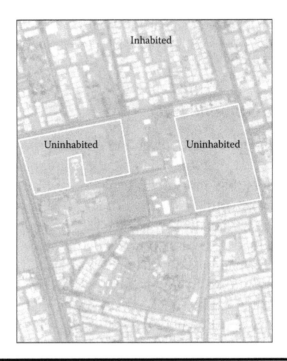

Figure 7.1   Example of uninhabited and inhabited areas for calculating proxy population distribution and density.

then be divided by the total inhabited area of each district, resulting in a better measurement of population density (Khormi and Kumar, 2011b).

Another socioeconomic factor that can be verified using satellite images is neighbourhood quality. This can be done based on several factors, such as the density of the buildings in each neighbourhood in each zone, the width of the streets, and the roof area of the buildings. In the absence of available detailed neighbourhood quality information for each zone, it can be decided that a factor based on these variables is the best-available option on the ground at the local level.

The first step is to determine the criteria that can be used for the classification of the neighbourhood qualities of each district. Generally, in any city or district, a number of variables determine city or district quality. For instance, the wealthier areas generally have wider streets and buildings that are

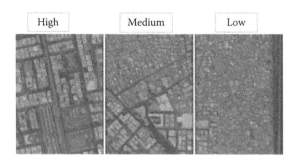

Figure 7.2 Three different neighbourhood qualities can often be found in each zone.

farther away from street curbs; the poorer areas generally have narrower streets and buildings that are closer to the street curbs. Thus, the distances from the edges of the buildings on the two sides of the streets are smaller in poorer neighbourhoods as compared to wealthier neighbourhoods. Also, in poorer neighbourhoods, the number of buildings on a given property is higher, the building sizes are smaller, and the buildings are closer together (Khormi and Kumar, 2011b). In wealthier neighbourhoods, the number of buildings on a given property is lower, the building sizes are bigger, and the buildings are widely separated (Khormi and Kumar, 2011b). For example, in studies conducted in Saudi Arabia, the estimated average size of blocks of land in poorer neighbourhoods was around 125 m², but it was greater than 280 m² in wealthier suburbs (Figure 7.2 and Table 7.2).

Regarding street width, we (Khormi and Kumar, 2011b) measured the distances between the edges of the buildings on opposite sides of the streets in a number of different areas. We found that the average distances between the edges of the buildings were between 1 and 5 m in the poorer neighbourhoods and were 10 m or wider in the wealthier neighbourhoods.

Building areas were measured using the roof area because it was difficult to measure the size of the blocks of land. We recognized that this technique would not account for

Table 7.2 Criteria Used to Classify Districts Based on Neighbourhood Quality

Building Size	Street Width	Number of Buildings per 0.25 km²
Low Class		
≤150 m²	<5 m	>400
Medium Class		
150–250 m²	5–10 m	200–400
High Class		
>250 m²	>10 m	<200

multilevel building construction, but in the area where this study was undertaken, there were few multilevel buildings.

Based on this information, a low-quality neighbourhood was defined as having a street width of 1 to 5 m, a building density of 400 buildings per 0.25 km² or greater, and an average roof area of 150 m² or less. A medium-quality neighbourhood was defined as having a street width of 5 to 10 m, a building density of between 200 and 400 buildings per 0.25 km², and an average roof area of between 150 and 250 m². A high-quality neighbourhood was defined as having a street width of 10 m or wider, a building density of 200 buildings or less per 0.25 km², and an average roof area of more than 250 m² (Table 7.2).

After the criteria were determined, the study area was divided into a number of neighbourhoods (e.g., 2000 neighbourhoods) using Hawth's tools. From the sampling tools, a polygon grid was created with 0.5 km as the spacing between lines, for example. This tool snapped the vector grid to a major coordinate system interval (as defined by the spacing between the lines).

Each neighbourhood was then assigned to one of the three quality classifications, according to the criteria in Table 7.1

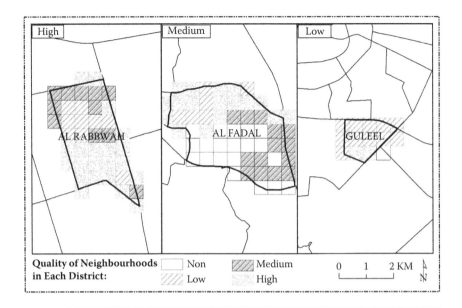

Figure 7.3   A method of assigning each district to a quality class.

(Figure 7.3), and an overall class was then assigned to each neighbourhood based on the majority of classifications in that neighbourhood. Then, an overall class was assigned to each district based on the majority of classifications of the neighbourhoods in that district (Figure 7.4).

7.3 Modelling Spatial Associations

7.3.1 Geographically Weighted Regression and Ordinary Least Squares

Identifying and measuring relationships help provide better understanding of disease spatial distribution. In addition, they help to predict where diseases are likely to occur. They are ways of examining causes of why diseases like DF occur where they do. One of the spatial techniques that helps in identifying and measuring spatial relationships is OLS. For all spatial regression analyses, OLS is the suitable initial point;

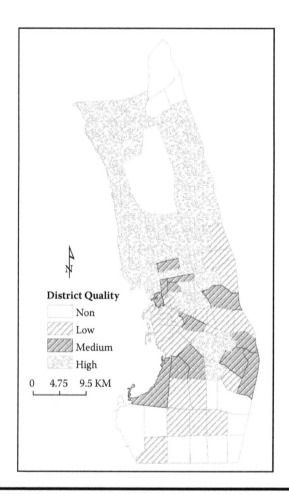

Figure 7.4   An overall class was assigned to each district based on the majority of classifications of the neighbourhoods in that district.

it offers a global model of the variables or processes that researchers or health officials try to understand or predict using a single regression equation to represent that process or variable (Khormi and Kumar, 2011b).

GWR describes a family of regression models that are allowed to vary spatially. As a target unit, GWR uses the longitude and latitude coordinates of each sample point (e.g., DF or malaria case locations, the district recorded number of cases, or the zone centroid as a form of spatially weighted least squares regression). The next paragraph explains the result of a model of this form.

Examining sets of DF case locations within well-defined neighbourhoods of each of the sample DF locations as points or district centroids helps to determine the coefficient $\beta(t)$. The well-defined neighbourhoods are, essentially, circles or radii r around case locations. However, if r is treated as a fixed value in which all points are regarded as having equal importance, it could include every point (for a large r) or, alternatively, no other points (for a very small r). Instead of using a fixed value for r, it is replaced by a distance-decay function, $f(d)$. This function may be finite or infinite, as with kernel density estimation.

The following is an example to provide an overview of how GWR can be implemented to model mosquito-borne disease (e.g., DF risks based on population number, population density, dengue cases, and neighbourhood quality are used). Before using this modelling method, the modifiable areal unit problems (MAUPs) need to be considered because they can arise from the fact that those areal units are usually arbitrarily determined and modifiable, in the sense that they can be aggregated to form units of different spatial arrangements. Disease data is sometimes combined into sets of increasingly larger areal units or alternative combinations or base units at equal or similar scales. Therefore, it is particularly important to use MAUPs when referring to variations in results (Openshaw and Taylor, 1979; Khormi and Kumar, 2011b). It is a common problem for disease surveillance and control programs if data are mostly reported for areal units (Scott and Morrison, 2003). However, in this example, the data of DF cases were collected and reported at the district level, and the data was neither combined nor aggregated, which minimized MAUP errors. Health-related data are seldom available on a per-person basis for confidentiality reasons, and the disease data was no different. For modelling purposes, the disease data was used at the finest resolution at which it was collected (Khormi and Kumar, 2011b).

Modelling associations between DF cases and socioeconomic factors, such as population, population density, and neighbourhood quality, is critical. There are a number of methods that can be used to model the spatial relationships, although two of the common spatial methods are GWR and OLS. Unreliable results are usually found when two or more variables are redundant, when they lump together, or when significant spatial autocorrelation exists among the data, especially in global regression models, such as OLS. Alternatively, a local regression equation for each feature (e.g., district) in the data set can be built using GWR (Nakaya et al., 2005; Spurna, 2008; Khormi and Kumar, 2011b). As a result, it is better to use GWR on a local scale because it is better able to handle spatial autocorrelations.

OLS and GWR should be examined to determine which method would better fit the observations. This can be determined through the results of the Akaike information criterion (AIC). The AIC resulting from GWR should be compared to the AIC resulting from OLS to determine which method would better fit the observed data. In this example, GWR provided a better fit to the observed data because the AIC of GWR ($\approx$1170) was lower than the AIC of OLS ($\approx$1200). It had the additional benefit of moving from a global model (OLS) to a local regression model (GWR).

GWR was used to describe this spatial association because it can measure the spatial dependency in a data set, and it is easily understood because of the traditional regression-based framework (Boots, 2003; O'Sullivan, 2003; Khormi and Kumar, 2011b). Before applying GWR, dependent and explanatory variables were determined. The total number of DF cases in each district was determined as the dependent variable, and the population, population density, and neighbourhood quality of each district were determined as the explanatory variables (Figure 7.5). The kernel was specified as a fixed distance to solve each regression analysis, and the bandwidth was specified using the AIC to determine the extent of the kernel.

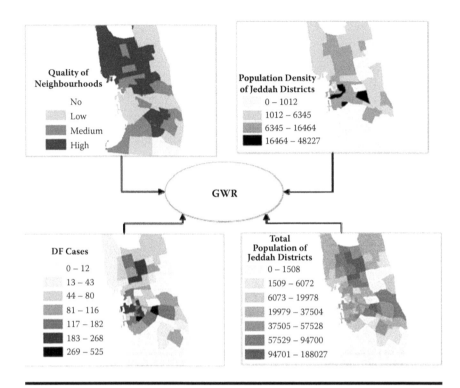

Figure 7.5   The framework of the human risk prediction model.

This was the bandwidth or number of neighbours used for each local estimation, and it was perhaps the most important parameter for the GWR because it controlled the degree of smoothing in the model. Predicted values that were used to build the model of humans at risk of DF were estimated by GWR. The example model used socioeconomic parameters, such as population, population density, and neighbourhood quality, to create a prediction model of the likelihood of humans becoming infected by DF in the districts of Jeddah (Figure 7.6). GWR revealed a strong positive association ($R^2 = 0.80$) between DF cases and population, population density, and neighbourhood quality in Jeddah. The model showed that a population of around 41% (≈1,173,701) is at a high risk of contracting DF; a population around 42% (≈1,212,733) is at a medium risk; and one around 17% (≈503,265) is at low risk for DF.

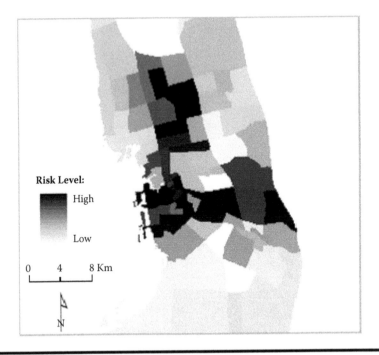

Figure 7.6 **Prediction model of humans at risk of DF based on different socioeconomic parameters.**

7.3.2 Linear and Multiple Regression

To model the spatial relationship between two variables (e.g., the mosquito and the disease case) by fitting a linear equation to the gathered data, linear regression can be used. The vector is considered an independent variable; the disease case is a dependent variable. Health officers or other modellers might want to relate the weights of individuals to their heights using a linear regression model.

The equation of the linear regression form is as follows:

$$Y = a + bX$$

where X is the explanatory variable (e.g., mosquito), and Y is the dependent variable (e.g., DF cases). The slope of the line is b, and a is the intercept (the value of y when $x = 0$).

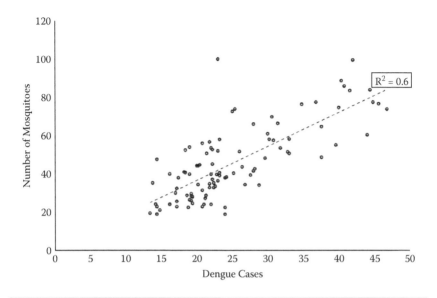

Figure 7.7   The association between the observed data for the DF cases and the number of mosquitoes.

The modeller must know whether there is a relationship between the variables of interest before fitting a linear model to the data. This does not necessarily imply that one variable causes the other (e.g., higher mosquito numbers do not cause higher disease cases), but that there is some significant association between the two variables. Fitting a linear regression model to the data that shows no relationship between the proposed explanatory variable, such as the mosquito, and the dependent variables, such as the DF cases, will not provide a useful model. A valuable numerical measure of association between the two variables is the correlation coefficient (R^2), which is a value between −1 and 1, indicating the strength of the association of the observed data for the two variables (Figure 7.7).

In reality, a number of variables often affect disease distribution. For example, temperature, relative humidity, and rainfall affect the abundance of the adult mosquitoes that transmit DF. Therefore, a multiple regression technique is powerful in predicting or forecasting the future value of a variable from the current value of two or more variables, such as minimum and maximum (min/max) temperatures and min/max relative humidity.

Predicting the weekly adult female *A. aegypti* population that transmits DF in advance based on meteorological variables can help public health officials to control mosquitoes and mosquito-borne diseases, such as DF (Hales et al., 2002). Knowing the amount of lag time between independent variables (e.g., meteorological factors) and dependent variables (e.g., *Aedes* mosquito abundances) is necessary to develop prediction models. The deposition of mosquito eggs, and their maturation into larvae and then into adults, requires aquatic breeding sites and time and is therefore dependent on meteorological factors and time (Aburas, 2007; Morrison et al., 2008).

An understanding of DF risk based on mosquito abundance and meteorological variables can be developed, applied sustainably, and measured prospectively. The development of this understanding is a priority for modelling risks of mosquito-borne diseases in general and DF specifically (Morrison et al., 2008). If disease risk factors could be included in a predictive model, health authorities would then have quantitative measures that could be used to instigate advanced vector control operations. This is an approach rarely used in disease control.

An example is shown here to examine the utility of various entomological indices to determine how best to create a predictive model of disease risk. In doing so, the predictive model will help health officers increase their ability to predict indices ahead of time, which would be useful for health authorities when planning control responses.

Before the multiple regression analysis, a Pearson's correlation analysis was used to verify the associations between adult female *A. aegypti* and meteorological variables. It is important to determine the variable that has the highest correlation with mosquitoes and to avoid multicollinearity. Pearson's correlation analysis is an option that can be used to measure associations; it depends on the assumption that both dependent and independent (*X* and *Y*) variables follow a normal distribution. It shows the degree of the linear relationship between two variables (Khormi et al., 2013). This helps in determining the highest correlations between the variables that will be used in the multiple regression analysis.

Calculating the weekly average number of adult *A. aegypti* mosquitoes, the weekly average min/max temperatures, and the weekly average min/max relative humidity from 2007 to 2010 were collected before the correlation analysis. Then the correlations between the average weekly mosquitoes and the weekly values of the selected meteorological variables at different weekly lags were calculated. In this example, eight time lags were used to avoid multicollinearity before running the multiple regression analysis (Figure 7.8). To reduce or eliminate multicollinearity, climatic variables at lag times that did not seem logically essential to the model were removed. In addition, increasing the sample size in the analysis and using 4-year daily data to obtain narrower confidence intervals

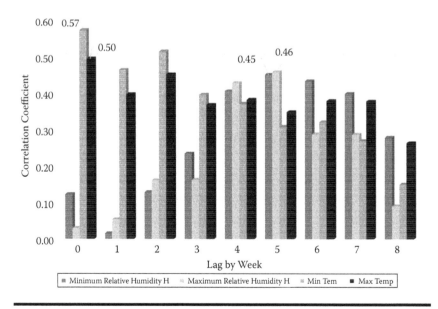

Figure 7.8 Correlation coefficients among climatic variables and mosquitoes, depending on different time lags.

despite multicollinearity were important for reducing the impact of collinearity (Khormi et al., 2013).

As Figure 7.8 shows, the highest correlation between temperatures and mosquitoes was found at lag 0 week; the highest correlation between relative humidity and mosquitoes was found at lag 5 week. The multiple regression analysis was undertaken using all correlated climatic variables at these time lags to create the first model. The two highest correlations were found between mosquitoes and the minimum temperature at lag 0 week and the maximum relative humidity at lag 5 week, which were then used to create the second model. They were used because they could better explain the differences in the average weekly mosquitoes. However, mosquito and meteorological data from 2007 to 2009 were used in the multiple regression analysis. Based on the result of this analysis, weekly mosquito values for 2010 were predicted. To assess models 1 and 2 for accuracy, the actual weekly female *A. aegypti* data from 2010 were used, and correlation models were performed between the predicted and actual values of the models.

The min/max relative humidity of lag 5 week and the min/max temperatures of lag 0 week were used in the first model (Table 7.3), and others were excluded because of their lack of significant correlation with mosquito numbers (Figure 7.8). There was a decrease in the average weekly minimum temperature of lag 0 week (the current week) prior to the mosquito collection week. This was considered the most significant entomological indicator for the rise in the weekly abundance of adult *A. aegypti*. Then, an increase in the weekly min/max relative humidity at 35 days (5 weeks) was found prior to the mosquito collection week. The average maximum temperature of the current week (lag 0) of collecting the mosquitoes seemed to have an effect on the increase in the average weekly amount of mosquitoes, but to a lesser extent when compared to the min/max relative humidity and, in a different direction, compared to the minimum temperature.

Table 7.3   Meteorological Variables with the Most Significant Correlations with the Average Number of Adult Female *Aedes aegypti* Mosquitoes (Model 1)

Variable	Coefficients	Standard Error	P Value	$CI_{95\%}$
MaxRH Lag 5 week	1.34	0.67	0.05	0.00 to 2.69
MinRH Lag 5 week	1.38	0.52	0.01	0.33 to 2.42
MaxTemp Lag 0 week	0.83	2.10	0.69	−3.39 to 5.04
MinTemp Lag 0 week	−3.43	2.39	0.16	−8.23 to 1.37

Note: Model 1: Average female *Aedes aegypti* = 1.34*MaxRH Lag 5 week + 1.38*MinRH Lag 5 week + 0.83*MaxTemp Lag 0 week + −3.43*MinTemp Lag 0 week

Table 7.4   Significant Meteorological Variables with the Highest Correlation with the Average Number of Adult Female *Aedes aegypti* Mosquitoes (Model 2)

Variable	Coefficients	Standard Error	P Value	$CI_{95\%}$
MinTemp Lag 0 week	−3.43	2.39	0.16	−8.23 to 1.37
MaxRH Lag 5 week	1.34	0.67	0.05	0.00 to 2.69

Note: Model 2: Average female *Aedes aegypti* = −3.43*MinTemp Lag 0 + 1.34*MaxRH Lag 5

For the analysis of the second model, only the minimum temperature of lag 0 week and the maximum relative humidity of lag 5 week were used because they were significantly related to the average number of mosquitoes (Table 7.4 and Figure 7.8).

From model 1, there was a resulting correlation coefficient of 0.60 observed between the actual and predicted average

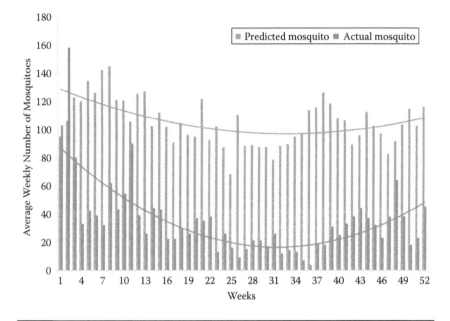

Figure 7.9 Predicted mosquito abundance based on model 1 that depended on four climatic variables.

mosquito numbers. It was—albeit with different scales—lower in actual mosquito numbers and higher in the predicted values (Figure 7.9). On the other hand, there was a higher resulting correlation coefficient of 0.69 observed between the actual and the predicted values with narrower bands (Figure 7.10). In terms of showing the trend of mosquito weekly abundance, model 2 predicted values that were more accurate compared to model 1 (Figures 7.9 and 7.10).

More immediate effects of weather on vector abundance can be seen in model 2. For this reason, it was selected as the better model compared to model 1. It was found to be sensitive in predicting sharp increases (e.g., from week 1 to week 2 and from week 51 to week 52) and the reduction in mosquito numbers in some weeks (e.g., week 2 to week 3 and week 43 to week 44). This model was able to explain the weekly trends of mosquito numbers, which was observable from around week 1 to week 52.

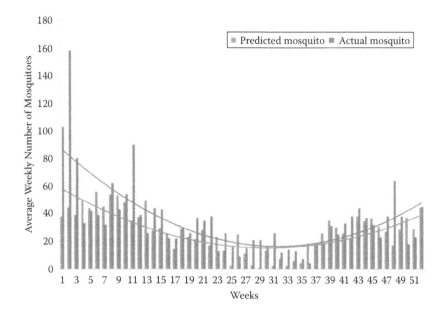

Figure 7.10   Predicted mosquito abundance based on model 2 that depended on two climatic variables.

The same method is implemented with the data of some districts to predict areas that will be under risk based on the predictive values of mosquito abundance that were determined from model 2 to improve monitoring and forecasting of mosquito abundance in each district or subdistrict of Jeddah (Figure 7.11). This method can provide a better understanding of future risk based on the climatic conditions in each area. These are the parameters used for modelling DF risks to better monitor and control these risks successfully.

Weather-based predictive modelling can be used to predict the weekly increases and decreases in vector abundance, which would allow vector control measures to be initiated before sharp increases happen (Hales et al., 2002). An easy method using multiple regression analysis is presented to predict the weekly abundance of mosquitoes. To build the predictive model, the forecasted minimum temperature and maximum relative humidity, which can be provided by climatic

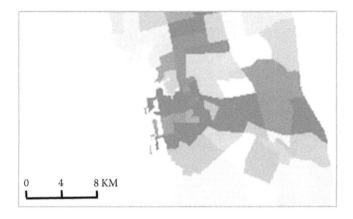

0   4   8 KM

Figure 7.11   Risk areas based on the climatic conditions in each area. The dark gray colour shows high-risk areas, and the light gray colour shows low-risk areas.

authorities, are needed. Based on this model, mosquito-borne disease control measures can benefit from these models to have an advanced plan for controlling the disease.

References

Aburas, H.M. (2007). ABURAS Index: A statistically developed index for dengue-transmitting vector population prediction. *Proceedings of World Academy of Science, Engineering and Technology,* 23:151–154.

Beck, L.R., Lobitz, B.M., Wood, B.L. (2000). Remote sensing and human health: new sensors and new opportunities. *Emerging Infectious Diseases,* 6(3): 217–227.

Boots, B. (2003). Geographically weighted regression: the analysis of spatially varying relationships. *International Journal of Geographical Information Science,* 17(7): 717–719.

Cromley, E.K., and McLafferty, S.L. (2012). *GIS and Public Health.* 2nd ed. New York: Guilford Publications.

Day, J.F., and Curtis, G.A. (1989). Influence of rainfall on *Culex nigripalpus* (Diptera, Culicidae) blood-feeding behaviour in Indian River County, Florida. *Annals of the Entomological Society of America,* 82(1): 32–37.

Dye, C., Reiter, P. (2000). Climate chance and malaria—temperatures without fevers? *Science,* 289(5485): 1697–1698.

Epstein, P.R. (2000). Is global warming harmful to health? *Scientific American,* 283(2): 50–57.

Friis, R.H., Sellers, T.A. (2009). *Epidemiology for Public Health Practice.* 4th ed.. Sudbury, MA: Jones and Bartlett.

Githeko, A.K., Lindsay, S.W., Confalonieri, U.E., Patz, J.A. (2000). Climate change and vector-borne diseases: a regional analysis. *Bulletin of the World Health Organization,* 78(9): 1136–1147.

Hakre, S., Masuoka, P., Vanzie, E., Roberts, D.R. (2004). Spatial correlations of mapped malaria rates with environmental factors in Belize, Central America. *International Journal of Health Geographics,* 3(6). doi:10.1186/1476-072X-3-6

Hales, S., de Wet, N., Maindonald, J., Woodward, A. (2002). Potential effect of population and climate changes on global distribution of dengue fever: an empirical model. *Lancet,* 360(9336): 830–834.

Khormi, H., Kumar, L. (2011a). Examples of using spatial information technologies for mapping and modelling mosquito-borne diseases based on environmental, climatic, socio-economic factors and different spatial statistics, temporal risk indices and spatial analysis: a review. *Journal of Food, Agriculture and Environment,* 9(2): 41–49.

Khormi, H., Kumar, L. (2011b). Modeling dengue fever risk based on socioeconomic parameters, nationality and age groups: GIS and remote sensing based case study. *Science of the Total Environment,* 409(22): 4713–4719.

Khormi, H.M., Kumar, L., Elzahrany, R.A. (2013). Regression model for predicting adult female *Aedes aegypti* based on meteorological variables: a case study of Jeddah, Saudi Arabia. *Journal of Earth Science Climate Change,* 5: 168. doi:10.4172/2157-7617.1000168

Koopman, J.S., Prevots, D.R., Vaca Marin, M.A., Gomez Dantes, H., Zarate Aquino, M.L., Longini, I.M. Jr., Sepulveda Amor, J. (1991). Determinants and predictors of dengue infection in Mexico. *American Journal of Epidemiology,* 3(11): 1168–1178.

Lindsay, S.W., Birley, M.H. (1996). Climate change and malaria transmission. *Annals of Tropical Medicine and Parasitology,* 90(6): 573–588.

Molyneux, D.H. (1997). Patterns of change in vector-borne diseases. *Annals of Tropical Medicine and Parasitology,* 91(7): 827–839.

Monath, T.P., Tsai, T.P. (1987). St-Louis encephalitis—lessons from the last decade. *American Journal of Tropical Medicine and Hygiene,* 37(3): 40–59.

Moore, C.G. (1985). Predicting *Aedes aegypti* abundance from climatologic data. In: Lounibos, L.P., Rey, J.R., Frank, J.H. (Eds.), *Ecology of Mosquitoes: Proceedings of a Workshop.* Vero Beach, FL: Vero Beach Medical Enomology Laboratory; 223–235.

Morrison, A.C., Zielinski-Gutierrez, E., Scott, T.W., Rosenberg, R. (2008). Defining challenges and proposing solutions for control of the virus vector *Aedes aegypti. PLoS Medicine,* 5(3): e68. doi:10.1371/journal.pmed.0050068

Nakaya, T., Fotheringham, A.S., Brunsdon, C., Charlton, M. (2005). Geographically weighted Poisson regression for disease association mapping. *Statistics in Medicine,* 24(17): 2695–2717. doi:10.1002/sim.2129

Openshaw, S., Taylor, P. (1979). *A Million or So Correlation Coefficients: Three Experiments on the Modifiable of a Real Unit Problem.* London: Pion.

O'Sullivan, D. (2003). Geographically weighted regression: the analysis of spatially varying relationships. *Geographical Analysis,* 35(3): 272–275.

Rakotomanana, F., Ratovonjato, J., Randremanana, R.V., et al. (2010). Geographical and environmental approaches to urban malaria in Antananarivo (Madagascar). *BMC Infectious Diseases,* 10: 173).

Scott, T.W., Morrison, A.C. (2003). Aedes aegypti *Density and the Risk of Dengue-Virus Transmission.* Davis: Department of Entomology, University of California, Davis.

Sithiprasasna, R., Ugsang, D.M., Honda, K., Jones, J.W., Singhasivanon, P. (2005). IKONOS-derived malaria transmission risk in northwestern Thailand. *Southeast Asian Journal of Tropical Medicine and Public Health,* 36(1): 14–22.

Spurna, P. (2008). Geographically weighted regression: method for analysing spatial non-stationarity of geographical phenomenon. *Geografie,* 113(2): 125–139.

Takken, W., Knols, B.G.J. (Eds.). (2007). *Emerging Pests and Vector-Borne Diseases in Europe: Ecology and Control of Vector-Borne Diseases.* Wageningen, the Netherlands: Wageningen Academic.

Troyo, A. (2007). Analysis of Dengue Fever and *Aedes aegypti* (Diptera: Culicidae) Larval Habitats in a Tropical Urban Environment of Costa Rica Using Geospatial and Mosquito Surveillance Technologies. PhD dissertation, University of Miami.

Wegbreit, J. (1997). The Possible Effects of Temperature and Precipitation on Dengue Morbidity in Trinidad and Tobago: A Retrospective Longitudinal Study. PhD dissertation, University of Michigan, Ann Arbor.

Chapter 8

Global Climate Change and Modelling the Potential Distribution of Vector-Borne Disease

8.1 Introduction

Climate plays an important role in disease vectors and in the replication and transmission of vector-borne diseases. It also circumscribes the distribution of diseases and has an impact on the timing and intensity of outbreaks (Wayne and Graham, 1968; Hopp and Foley, 2001).

The average global temperature may increase by between 2.0°C and 6°C from 1990 to 2100, as suggested by the Intergovernmental Panel on Climate Change (IPCC; 2007). Minimum temperatures are now increasing at a much faster rate than predicted (IPCC, 2007), and this is changing disease distribution. This increase is predicted to continue under the anticipated climate change scenarios. As a result, diseases and their vectors are likely to be extended into regions previously free of the disease, or such conditions may exacerbate transmission in the parts of the world to which they are endemic (IPCC, 2007). Therefore, an increase in temperature and other associated climate changes may affect and modify the geographical distribution and range of vector-borne diseases in the future (Hopp and Foley, 2001; Johansson et al., 2009).

There is increasing scientific interest in the potential effects of global climate change on health. Global climate change re-emerged recently and spread mosquito-borne diseases to new areas, extending the risk season for infection and maintaining a high incidence level. The spread of global warming has led to the spread of mosquito-borne diseases to new suitable habitats, such as water pools. Dengue fever (DF), Lyme disease, malaria, West Nile virus, and yellow fever are examples of mosquito-borne diseases that are influenced by climate change (Takken and Knols, 2007).

Rapid climate change (e.g., sustained global warming of 0.2°C per decade) raises many questions about how mosquito-borne diseases will be affected. many studies have recently been conducted to answer these questions. For example, Hales et al. (2002) modelled the reported global distribution of DF based on vapour pressure, which is a measure of humidity. They also assessed the changes in the geographical limits of DF transmission and in the number of people at risk of DF by incorporating future climate change and human population projections into the model. Estimates of population and climate change projections for 2085 showed that 5–6 billion people would be at risk of DF transmission, compared with 3.5 billion people if climate change did not happen (Hales et al., 2002).

In this chapter, we show how to model and project the related potential alterations in climate change on the distribution of *Aedes aegypti* based on different climatic factors of temperature, moisture, dryness, and heat. The same techniques can be used for other vector-borne diseases. We use different climate models (CMs) and scenarios available in CLIMEX software to model the future distributions. Modelling the future distributions is dependent on the climate responses of *A. aegypti* mosquitoes and their current and invasive distributions in South and Southeast Asia. We then model and project their potential distribution under the current climate, using extensive distribution data for model validation, and assess the impacts of climate change on potential distribution using two CMs: CSIRO-Mk3.0 (CS) and MIROC-H (MR). These were run with A2 SRES (Special Report on Emissions Scenarios) for 2050 and 2100.

The *Aedes* mosquito, in relation to dengue, has been described since 1902 in some South and Southeast Asian countries, especially when Malaysia and Singapore had experienced a series of dengue epidemics that affected most of the port cities from 1900 to 1901 (Skae, 1902; Siler et al., 1926; Rudnick, 1986). The outbreaks were correlated then with *A. aegypti* distribution (Siler et al., 1926; Gubler, 1997). These outbreaks and distributions occurred only in the seaports and along parts of the seacoast and were not recorded inland. Previously, some studies had shown that dengue and *A. aegypti* were recorded and were widespread in rural and urban areas and near areas of human habitation in this region (Smith, 1956; Skae, 1902; Graham et al., 1999).

The South and Southeast Asia region is one of the tropical and subtropical areas that have been affected by dengue, which is transmitted mainly by *A. aegypti,* creating a public health problem in the region (Halstead, 2006; Graham et al., 1999). DF has been considered one of the leading causes of hospitalization in South and Southeast Asia and is the cause of thousands of infections and hundreds of deaths. In this region, dengue causes a substantial health burden on both the health care system and individual households (Clark et al., 2005; Rogers et al., 2006; Graham et al., 1999).

Most of the South and Southeast Asian countries are situated in the tropical and subtropical regions that have warmer temperatures. The warmer temperature increases the disease virus and vector development rates, which can raise the odds of dengue transmission (Khormi and Kumar, 2011; Khormi et al., 2011).

Many studies show that climatic variables (e.g., temperature and rainfall) relate strongly to the estimated risk of dengue infection (Patz et al., 1998; Jetten and Focks, 1997; Chakravarti and Kumaria, 2005; Khormi et al., 2011). Currently, the potential effects of climate change on health generally and dengue particularly are receiving a great deal of scientific interest (Haines et al., 2006). Climate change merged and spread dengue and its vector to new areas, extending the risk for infection and maintaining a high incidence level (Takken and Knols, 2007). For example, the *Aedes* mosquito was formerly observed in elevations around 1500 m and currently observed in areas above 2200 m because of climate change. These changes were caused by an increase in temperature (Suarez and Nelson, 1981; Herrera-Basto et al., 1992).

8.2 Distribution of Dengue Disease and Its Vector

As an example, we selected South and Southeast Asia as this region has some of the most densely populated areas of the world, contains about 49% of the world's population, has some of the highest population growth rates, and has large areas currently infested by *A. aegypti* or environmentally suitable for infestation by these mosquitoes (Gubler and Clark, 1995). The projected population of this region is over 4 billion by 2050 and over 3.7 billion by 2100 (United Nations, 2004).

Figure 8.1 shows the countries included in this example as well as the current approximate distribution of *A. aegypti* mosquitoes. Information on the distribution of *A. aegypti* in South and Southeast Asia was collected from different sources, such as the Centers for Disease Control and Prevention (CDC)

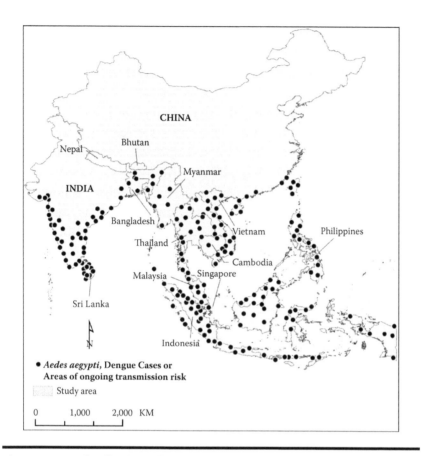

Figure 8.1   Distribution of the *Aedes aegypti* mosquito, dengue cases, and areas of ongoing transmission risk in the study area.

(2012); the Center for International Earth Science Information Network (CIESIN) (2012); Gratz (2004); Gubler and Trent (1994); Gubler and Clark (1995); Gubler (1998a, 1998b, 2002, 2003, 2004); Gubler and Kuno (1997); Lounibos (2002); Medlock et al. (2005); Moore (1999); Moore and Mitchell (1997); Malavige et al. (2004); Ooi et al. (2006); Li et al. (1985); Lee and Rohani (2005); Graham et al. (1999); Sumarmo et al. (1983); and Corwin et al. (2001). The identified locations show areas of ongoing transmission risk and local and regional dengue or imported cases of dengue and *A. aegypti* and are based on data from the ministries of health, international health organizations, journals, and knowledgeable experts (see references in this paragraph).

According to Gubler (1997), it is generally accepted that areas plagued by DF are also infested by *A. aegypti* because this is the main vector transmitting the disease. Figure 8.1 shows that most of the *A. aegypti* and dengue cases are recorded in many places in India, such as Thiruvananthapuram, Kerala, Madurai, Tamil Nādu, Mangalore, Karnātaka Chikungunya, Chennai, Bengaluru, Ganjām, Orissa, Mumbai, Maharashtra, Kolkata, West Bengal, Guwahati, Assam, Visakhapatnam, East Siang, and Arunāchal Pradesh. In Vietnam, Myanmar, and Bhutan, dengue and its vector have been recorded in areas of Bên Tre Province Dong Nai, Bình Thuận Province, Mon State, Kayin State, Phuentsholing, Chukha, Norgay, and Dega. In Thailand, dengue and its vector are mostly recorded in areas of Mae Hong Son, Nakhon Ratchasima, Lampang, and Buriram. In Cambodia, south of China and Taiwan, the disease is found in Foshan, Guangzhou, Guangdong Province, Kaohsiung City, Tainan City, and Taiwan. In the Philippines, they have been recorded, for example, in Manila, Luzon, Calabarzon region, Davao, and the Province of Negros Oriental. In Malaysia and Indonesia, dengue and the *Aedes* mosquito have been recorded in areas such as Selangor, Kuala Lumpur, Peninsula, Johor State, Lundu district in Sarawak, Yogyakarta, southern Sumatra, Jakarta, Bandung, and West Java.

8.3 CLIMEX, Climatic Models, and Data

CLIMEX for Windows (Version 3) was used to model the current and future potential distribution of *A. aegypti* under current and future climate scenarios, respectively. CLIMEX software allows users to model the potential distribution of a species by incorporating a variety of information types, including direct experimental observations of a species in regard to the growth response to environmental variables, its phenology, and knowledge of its spatial distribution (Wharton and Kriticos, 2004; Taylor et al., 2012a). It is based on the premise that the distribution envelope of plants and poikilothermal animals is primarily determined by climate (Andrewartha and Birch, 1954; Poutsma et al., 2008; Sutherst and Bourne, 2009; Chejara et al., 2010; Taylor et al., 2012a). The software is based on the formation of an ecophysiological model that assumes that species will flourish during a favourable season, resulting in positive population growth, and will not do so well during an unfavourable season, resulting in population decline (Sutherst et al., 2007).

CLIMEX uses the input data on current distributions and climatic conditions to generate an annual growth index (GIA) that gives the potential for population growth or decline during favourable and unfavourable conditions, respectively. Four stress indices (cold, wet, hot, and dry) and up to four interaction stresses (hot–dry, hot–wet, cold–dry, and cold–wet) are used to describe the probability that the population can survive unfavourable conditions (Taylor et al., 2012a,b). The weekly growth and stress indices are combined to form the Ecoclimatic Index (EI), which is an index of overall annual climatic suitability and is theoretically scaled from 0 to 100. The species being modelled will only establish if the EI is greater than 0. Generally, EI values close to 100 (maximum) are rare and usually are confined to species with an equatorial range, as this would imply ideal growing conditions during the whole year (Sutherst et al., 2007; Taylor et al., 2012a,b). When EI values are close to zero, they indicate a low probability of conditions conducive to persistence in time and space.

A number of CMs exist for characterizing future climate (IPCC, 2007). For this example, we used the CS and the MR models. According to the MR model, temperature will increase by approximately 4.31°C by 2100; the CS model predicts a rise of 2.11°C during the same period. The two models also differ in terms of rainfall predictions, with the CS model predicting a 14% decrease in future mean annual rainfall, and the MR model predicting a 1% decrease (Chiew et al., 2009, Suppiah et al., 2007). We selected these two CMs because the temperature, precipitation, mean sea-level pressure, and specific humidity variables were available for both models in the CliMond data set (IPCC, 2000; Gordon et al., 2002) that was used in our modelling. Also, both these models have reasonably small horizontal grid spacings, and they perform well compared to other CMs based on observed climate at regional scales (Taylor et al., 2012a,b).

The A2 SRES scenario was selected to characterize one of the possible climate scenarios in the future. According to Kriticos et al. (2011) and Taylor et al. (2012a), the A2 scenarios consider a world with high population growth, slow economic development, and technological change. They assume neither very high nor low global greenhouse gas (GHG) emissions compared to the other scenarios, such as A1F1, A1B, B2, A1T, and B1 by 2100 (Suppiah et al., 2007). No scenarios from the B family of SRES were included in this research as observations showed that some parameters, such as global temperature and sea-level rise, are currently increasing at a much greater rate than predicted by the hottest B family SRES (Rahmstorf et al., 2007).

The CliMond 10' gridded climate data (Kriticos et al., 2011) was used for modelling. Average minimum monthly temperature T_{min}, average maximum monthly temperature T_{max}, average monthly precipitation P_{total}, and relative humidity at 0900 h ($RH_{09:00}$) and 1500 h ($RH_{15:00}$) were used to represent historical climate (averaging period 1950–2000). The same five variables were used to characterize potential future climate. We used the two CMs and the A2 scenario to identify suitable climate envelopes for *A. aegypti* distributions in 2050 and 2100.

We used distribution data of both *A. aegypti* and some dengue cases (if there was a lack of *A. aegypti* data), temperature and moisture indices, and cold and dry stresses to fit the CIMEX parameters as we felt these data could produce a model that better approximated the potential distribution. All the parameters were fitted to the known and naturalized distribution of *A. aegypti*. Parameters were adjusted iteratively until there was reasonable agreement with the current *Aedes* distribution, and the fitted parameters were checked to ensure that they were reasonable.

A number of studies have suggested that *Aedes* mosquitoes are critically dependent on temperature, and that temperatures from 14°C to 18°C at the lower end and from 35°C to 40°C at the upper end could lead to higher transmission occurrence (Wallis, 2005). Khormi et al. (2011) found that a minimum temperature range from 18°C to 25°C was suitable for mosquito survival in Jeddah, Saudi Arabia, and that the *Aedes* mosquito survival rate increased at higher temperatures (but at not more than 38°C). Wayne and Graham (1968) and Connor (1924) found that *A. aegypti* was most active within the temperature range between 15°C and 30°C. Other field observations and laboratory tests have determined the survival rates to be from about 18°C to not more than 38°C, based on daily or monthly data of minimum and maximum temperatures (Macfie, 1920; Bliss and Gill, 1933; Christophers, 1960). Therefore, we set the limiting low temperature DV0 at 18°C, the lower optimal temperature DV1 at 25°C, the upper optimal temperature DV2 at 32°C, and the limiting high temperature DV3 at 38°C. These sets provided the best fit to the observed distribution of *A. aegypti*.

The lowest limiting moisture SM0 was set at 0 because it represents the permanent wilting point, and this number provided a good fit with the observed distribution of *A. aegypti* and dengue in drought areas, particularly in some areas of Indochina (e.g., Thailand, Vietnam, Myanmar, Cambodia, and Laos). The lower (SM1) and upper (SM2) optimum moisture and the highest limiting moisture were set at 0.2, 0.5, and 4, respectively, if species growth was to be increased in countries such as Malaysia and Indonesia. In addition, these values provided an appropriate fit to the observed distributions.

The heat stress parameter TTHS was set at 38°C because it is reported that, in some countries, *A. aegypti* is able to survive up to this temperature (Khormi et al., 2011; Gubler and

Clark, 1995). The heat stress accumulation rate THHS was set at 0.9 week^{-1}, which allows *A. aegypti* to persist along most of the Southeast Asian islands.

The dry stress parameter was set at 0.001 for the dry stress threshold SMDS and −0.001 week^{-1} for the dry stress rate HDS because these adjusted values provided an appropriate fit to the observed distribution.

The output of the model is the EI, which usually ranges from 0 to 100. For more efficient visualization of the potential and future distributions around the world, we transferred the results of the EI to Tahdeed. Tahdeed is a software that was developed by GIS Technology Innovation Centre, Saudi Arabia, to model risks of vector-borne diseases. The EI can only be converted as a point feature, each point having an EI value. We then converted the point feature to a raster and used the EI values to assign values to the output raster. Raster surfaces are effective in identifying where favourable climate areas are concentrated by highlighting areas based on the EI resulting from the CLIMEX analysis.

8.4 The Current and Potential Future Distribution

Figure 8.2 shows the modelled current distribution of suitable climatic conditions for *A. aegypti* based on point data (Figure 8.1) and our selected parameters (heat, wet, dry, and temperature indices) as obtained from CLIMEX. Comparing the modelled suitable climate (Figure 8.2) with the current recorded distribution of *A. aegypti* (Figure 8.1) shows that the present distribution of *Aedes* is consistent with high EI values resulting from the CLIMEX model. This shows that the selected values for the parameters used are optimum or close to optimum because there is a good match between the recorded

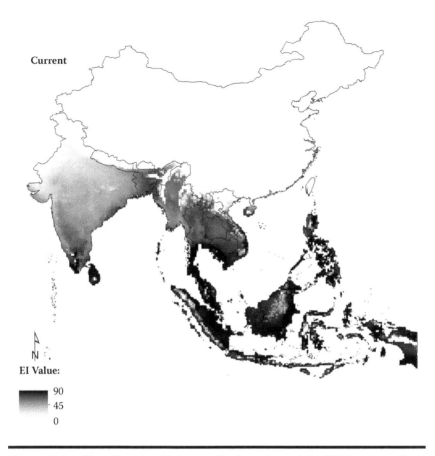

Current

EI Value:

90
45
0

Figure 8.2   The climate (EI) for *Aedes* based on CLIMEX for the refer-ence climate (averaging period 1950–2000). Favourability is based on the EI scale from 0 to 100. The unfavourable areas (EI = 0) are white, the marginal climate areas (EI = 1–10) are from light gray to dark gray, the favourable climate areas (EI = 10–20) are dark gray, and the very favourable areas (EI > 30) are from dark gray to black.

cases and modelled climate. Much of South and Southeast Asia are modelled to have favourable climatic conditions for *Aedes*. Sri Lanka, Malaysia, Indonesia, Singapore, Philippines, the southern coast of India, South Vietnam, Cambodia, and south-ern Thailand have high EI values and hence have suitable climatic conditions for the *Aedes* mosquito. Central and north-eastern India, Myanmar, northern Thailand, and Bangladesh have medium EI values, meaning they have suitable climatic conditions for *Aedes*. Most of China, Taiwan, Nepal, and Bhutan have very low EI values, so they do not have a conducive environment for *Aedes* survival.

The projected suitable climates for *Aedes* for 2050 under both CMs (CS and MR) are shown in Figure 8.3. There is a projected increase in climate suitability in Laos (mainly northern Laos), North Vietnam, Hainan Island in China, eastern Borneo and the Malaysia border region, southern Taiwan, the Assam region in India and in Sumatra, Indonesia. The regions where the climate suitability for *Aedes* actually decreases are central India, northern Thailand, and Cambodia. It should be noted that although the suitability (EI values) decreases, most of the area is still suitable for *Aedes* mosquitoes. For Myanmar, there are regions where the suitability increases (mainly eastern regions), but for some other areas (south and central), the suitability decreases. For China, new suitable areas evolve near the southern border close to Hong Kong, Macau, Zhanjiang, and south of Yunnan. However, most of China is still unsuitable for *Aedes*.

The projection of the CS model and the MR model for 2050 is similar. Both predict increases and decreases in the same regions, as discussed previously, although the levels of increase and decrease are slightly different. The CS model predicts larger decreases in suitability in India compared to the MR model.

Figure 8.4 shows the projected suitable climates for both models for 2100. The trend is similar to the models of 2050. There is an increase in suitability and suitable area extents in some regions while there is a decrease in suitability and suitable area extents in other regions. The regions where climate suitability increases (higher EI values) are Laos, North Vietnam, eastern Borneo, southern Taiwan, the Assam region in India, northern Myanmar, and southern regions of China. In China, we see large areas becoming conducive to *Aedes* survival, especially Hainan Island, areas near the southern border

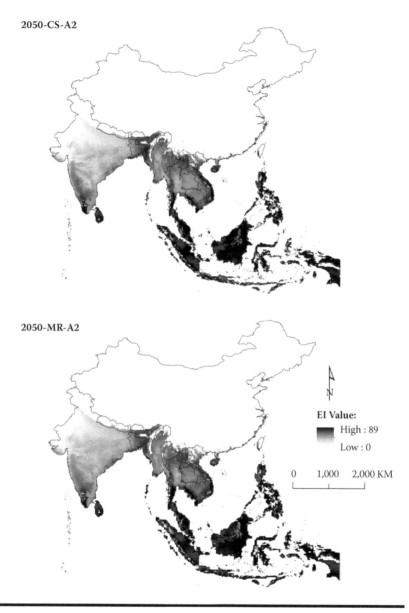

Figure 8.3 The climate (EI) for *Aedes* in 2050 based on CLIMEX under the CSIRO-Mk3.0 and MIROC-H GCMs running the A2 SRES. Favourability is based on the EI scale from 0 to 100. The unfavourable areas (EI = 0) are white, the marginal climate areas (EI = 1–10) are from light gray to dark gray, the favourable climate areas (EI = 10–20) are dark gray, and the very favourable areas (EI > 30) are from dark gray to black.

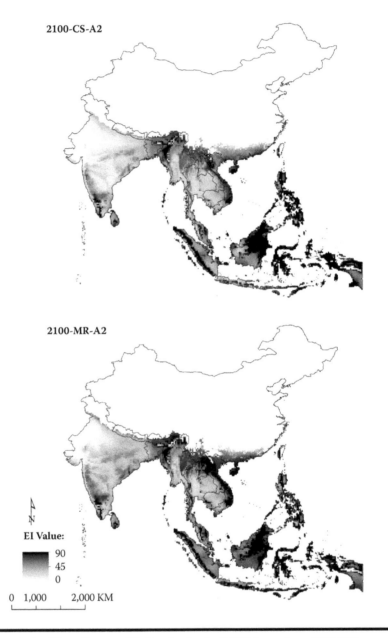

Figure 8.4 The climate (EI) for Aedes in 2100 based on CLIMEX under the CSIRO-Mk3.0 and MIROC-H GCMs running the A2 SRES. Favourability is based on the EI scale from 0 to 100. The unfavourable areas (EI = 0) are white, the marginal climate areas (EI = 1–10) are from light gray to dark gray, the favourable climate areas (EI = 10–20) are dark gray, and the very favourable areas (EI > 30) are from dark gray to black.

close to Hong Kong, Macau, south of Yunnan, and regions west of Guangzhou all the way to the Vietnamese border.

The regions where climate suitability decreases are large parts of India, Sri Lanka, Thailand, Cambodia, South Vietnam, southern Myanmar, western Borneo, Java, Sumatra, and Malaysia. The largest decreases are seen in India, Thailand, and Cambodia.

Both the CS and the MR models have similar predictions, with increases and decreases in similar regions. There are some differences in the levels of increase or decrease in suitability index. The MR model predicts a lower decrease in suitability in Indonesia, Malaysia, and central and eastern India than the CS model. The MR model predicts greater increases in suitability in Laos, eastern Myanmar, and northern Vietnam compared to the CS model.

8.5 Discussing the Potential Future Distribution

Many climatologists believe in the reality of climate change, and if this occurs, there is the potential for an increase in dengue-carrying mosquitoes transmitting the virus to susceptible human populations in the future to areas where currently there is no infestation. We have attempted to model and project the suitable climatic conditions and likely areas for *A. aegypti* in South and Southeast Asia using current and potential future climatic scenarios by using CLIMEX. Our current model shows a good fit with the current distribution records of *Aedes* and dengue, which were gathered for model validation purposes. The model shows Sri Lanka, Malaysia, Indonesia, Singapore, Philippines, the southern coast of India, South Vietnam, Cambodia, and southern Thailand having suitable to highly suitable conditions for the mosquito to survive and for dengue transmission. Some parts of central India are modelled to have suitable climatic conditions for *A. aegypti* and dengue transmission, exceeding the current known distribution for that area; this could be because of a lack of records from this region, or other nonclimatic factors, such as a lack of dispersal opportunities, could also inhibit *A. aegypti* from spreading in these regions. The mosquito distribution has limited establishment in much of China, northern India, Bhutan, Nepal, and northern Taiwan because of cold and dry stresses.

The model results also give an overview about the likely areas to be infested by *A. aegypti* in the future. Some areas where *Aedes* currently occur are projected to become climatically less suitable in the future. Figures 8.3 and 8.4 (scenarios for 2050 and 2100) indicate an overall reduction in the climate suitability for *Aedes* in many parts of South and Southeast Asia. Some of these reduced potential suitable regions for *A. aegypti* cover currently important hot spots, such as Sri Lanka, northern Thailand, Cambodia, and Central India.

The reason for this decline in suitability in the regions mentioned is changes in heat stress, causing large areas to have heat stresses beyond the maximum survival values for *Aedes*. For example, large areas in central to southern India that were below the heat stress limits become too hot for *Aedes* survival, with the heat stress increasing to beyond the maximum value by 2100 (Figure 8.5). The same applies for southern Myanmar, northern Thailand, Cambodia, and southern Laos. These regions have high suitability at present (Figure 8.2), but the suitability decreases markedly, and large areas in these regions become less suitable or unsuitable by 2100 (Figure 8.4).

These results highlight areas where climate suitability is expected to be decreased in the future, which is useful in making informed choices about the allocation of resources for mosquito control.

New areas of South and Southeast Asia that may be at risk of *A. aegypti* infestation under future climate conditions or where suitability increases are northern Laos, North Vietnam, eastern Borneo, southern Taiwan, the Assam region in India, northern Myanmar, and southern regions of China. Large areas

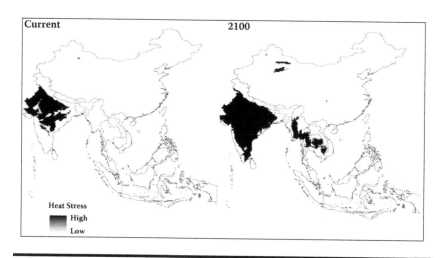

Figure 8.5   Changes in heat stress for *Aedes aegypti* **by 2100. Large areas of India, southern Myanmar, northern Thailand, Cambodia, and southern Laos become climatically less suitable because of increasing heat stress.**

of southern China that are currently unsuitable or marginal (Figure 8.2) become suitable for *Aedes* survival (Figure 8.4). The reason for this increase in suitability is the decreased impact of cold stress (Figure 8.6). Currently, most of China has medium-to-high cold stress levels (Figure 8.6), making the

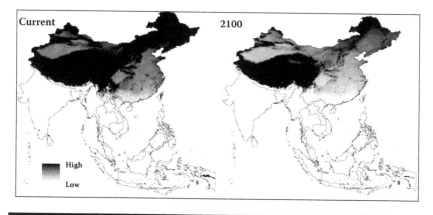

Figure 8.6   Changes in cold stress for *Aedes aegypti* **by 2100. Large areas of southern China and northern Vietnam, which currently have medium to high cold stress, are projected to have decreased cold stress values, hence making these regions more suitable for** *Aedes* **mosquitoes.**

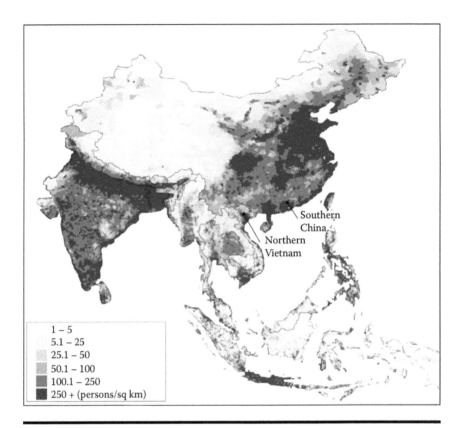

Figure 8.7 Population density for South and Southeast Asia.

survival of *Aedes* difficult. By 2100, it is projected that large areas of southern China will move from being below the critical cold stress value to within the suitable range for *Aedes* survival. Large areas in southern China will become highly suitable for *Aedes* infestation. These highly suitable areas have high population densities (Figure 8.7), greatly increasing the number of people who could be exposed to the *Aedes* mosquito and hence be susceptible to dengue infection. The same is true for northern Vietnam, where the new highly suitable regions are densely populated.

These changes in the suitability may warrant strategic control measures to prevent spread. Those areas may need more detailed risk evaluation regarding the spread of mosquitoes, and the assessment and management of mosquito risk depends to a large extent on projections of habitat suitability so the threat levels can be assessed.

A fundamental part of such assessments can be formed from the response of *A. aegypti* to changes in climate, especially in areas that are currently at risk and that will continue to be at risk from the mosquito in the future. Monitoring of these areas could be important. Health managers can use these projections in the decision-making processes and prioritize areas for eradication. Also, they can use them to determine areas where containment would be cost effective. Important information can be provided by computer simulation modelling in comparison to the traditional methods of investigation and to the management of complicated disease systems. Computer simulation modelling is helpful in assessing long-term variability in climate, for which historical studies lack similarity and prospective studies lack feasibility (Martens et al., 1995; Patz et al., 1998).

Areas that are identified currently with the dengue transmission intensity and with increases in this intensity in the future may be at more risk from haemorrhagic dengue. Other factors, such as the source of infection, the mosquito population size, and a susceptible human population, would also need to be present for a dengue epidemic to occur under suitable climatic conditions. Where and when dengue occurs in the future will depend on a number of different economic, social, environmental, and etiologic factors, such as urbanization, population density, poverty, inadequate mosquito control, absence of water systems, and international travel or migration, which are not addressed at this level of integrated modelling and should ultimately be incorporated to determine human risk from DF and its vector (Khormi and Kumar, 2011; Gubler and Clark, 1995). Future integrated models should attempt to account for these site-specific factors as well (Lifson, 1996) while climatic conditions contribute to epidemic spread and geographic distribution of dengue (Reiter, 1988; Reiter and Gubler, 1997).

The current model can be a useful tool in public aware-ness campaigns to enlist the help of health authorities in the management of existing infestations and in the prevention of further mosquito dispersal. Effectual mosquito monitoring systems and public education campaigns in this region could contribute toward more effective management.

8.6 Conclusion

The models presented here can help to reduce mosquito risk in South and Southeast Asian countries by improving early detection of mosquito-favourable areas to prevent DF preva-lence in the future. The strategy presented here should be supported by a sufficient surveillance system for viral diseases, which is now being implemented on a regional basis. These models can be used to manage and prevent the disease vector from extending into new areas and prioritize dengue manage-ment initiatives in current risk areas and those with continu-ing risk in the future. Health decision makers can adopt these models to make decisions regarding allocating resources for dengue risk toward areas where risk infection remains and away from areas where climatic suitability is likely to decrease under future scenarios.

This study has not attempted to explain all possible eco-nomic and social causes related to the distribution of mos-quito because we are not able to predict their status due to the complexity of their variables. Instead, we examined the relation between climate variability and the *Aedes* mosquito independently from other economic, environmental, etiologi-cal, and social variables. This study makes available valuable information for health policy makers who want to understand the potential effects of climate change on mosquito abundance to set priorities for mitigation and adaptation.

References

Andrewartha, H.G., Birch, L.C. (1954). *The Distribution and Abundance of Animals.* Chicago: University of Chicago Press.

Bliss, A.R., Gill, J.M. (1933). The effects of freezing on the larvae of *Aedes aegypti. American Journal of Tropical Medicine and Hygiene,* 13: 583–588.

Center for International Earth Science Information Network (CIESIN). Retrieved from http://www.ciesin.org/docs/001–613/map15.gif (accessed July 2012).

Centers for Disease Control and Prevention (CDC). (2012). DengueMap. Retrieved from http://www.healthmap.org/dengue/index.php (accessed July 2012).

Chakravarti, A., Kumaria, R. (2005). Eco-epidemiological analysis of dengue infection during an outbreak of dengue fever, India. *Virology Journal,* 2(32): 1–7.

Chejara, V.K., Kriticos, D.J., Kristiansen, P., Sindel, B.M., Whalley, R.D.B., Nadolny, C. (2010). The current and future potential geographical distribution of *Hyparrhenia hirta. Weed Research,* 50, 174–184.

Chiew, F.H.S., Kirono, D.G.C., et al. (2009). Assessment of rainfall simulations from global climate models and implications for climate change impact on runoff studies. 18th Annual World IMACS/MODSIM Congress, Ciarns, Australia, July 13–17. Available at http://mssanz.org.au/modsim09

Christophers, S.R. (1960). Aedes aegypti *(L.): The Yellow Fever Mosquito. Its Life History, Bionomics and Structure.* Cambridge, UK: Cambridge University Press.

Clark, D.V., Mammen, M.P., et al. (2005). Economic impact of dengue fever/dengue hemorrhagic fever in Thailand at the family and population levels. *American Journal of Tropical Medicine and Hygiene,* 72(6): 786–791.

Connor, M.E. (1924). Suggestions for developing a campaign to control yellow fever. *American Journal of Tropical Medicine,* 4: 277–307.

Corwin, A.L., Larasati, R.P., Bangs, M.J., et al. (2001). Epidemic dengue transmission in southern Sumatra, Indonesia. *Transactions of the Royal Society of Tropical Medicine and Hygiene,* 95: 257–265.

Gordon, H.B., Rotstayn, L.D., McGregor, J.L., Dix, M.R., Kowalczyk, E.A., O'Farrell, S.P., Waterman, L.J., Hirst, A.C., Wilson, S.G., Collier, M.A., Watterson, I.G., Elliott, T.I. (2002). *The CSIRO Mk3 Climate System Model.* CSIRO Atmospheric Research technical paper No. 60. Aspendale, Australia: CSIRO Atmospheric Research.

Graham, R.R., Juffrie, M., Tan, R., Hayes, C.G., Laksono, I., Ma'roef, C., et al. (1999). A prospective seroepidemiologic study on dengue in children four to nine years of age in Yogyakarta, Indonesia I. Studies in 1995–1996. *American Journal of Tropical Medicine and Hygiene,* 61(3): 412–419.

Gratz, N.G. (2004). Critical review of the vector status of *Aedes albopictus. Medical and Veterinary Entomology,* 18: 215–227.

Gubler, D.J. (1997). Dengue and dengue hemorrhagic fever: its history and resurgence as a global public health problem. In: Gubler, D.J., Kuno, G. (Eds.), *Dengue and Dengue Hemorrhagic Fever.* Wallingford, UK: CAB International; 1–22.

Gubler, D.J. (1998a). Dengue and dengue hemorrhagic fever. *Clinical Microbiology Review,* 11: 480–496.

Gubler, D.J. (1998b). Resurgent vector-borne diseases as a global health problem. *Emerging Infectious Diseases,* 4: 442–450.

Gubler, D.J. (2002). The global emergence/resurgence of arboviral diseases as public health problems. *Archives of Medical Research,* 33: 330–342.

Gubler, D.J. (2003). *Aedes albopictus* in Africa. *Lancet,* 3: 751–752.

Gubler, D.J. (2004). The changing epidemiology of yellow fever and dengue 1900 to 2003: full circle? *Comparative Immunology, Microbiology and Infectious Diseases,* 27: 319–330.

Gubler, D.J., Clark, G.G. (1995). Dengue/dengue hemorrhagic fever: the emergence of a global health problem. *Emerging Infectious Diseases,* 1: 55–57.

Gubler, D.J., Kuno, G. (1997). *Dengue and Dengue Hemorrhagic Fever.* Wallingford, UK: CAB International.

Gubler, D.J., Trent, D.W. (1994). Emergence of epidemic dengue/ dengue hemorrhagic fever as a public health problem in the Americas. *Infectious Agents and Disease,* 2: 383–393.

Haines, A., Kovats, R.S., Campbell-Lendrum, D., Corvalan, C. 2006. Climate change and human health: impacts, vulnerability, and public health. *Public Health,* 120: 585–596.

Hales, S., de Wet, N., Maindonald, J., Woodward, A. (2002). Potential effect of population and climate changes on global distribution of dengue fever: an empirical model. *Lancet,* 360(9336): 830–834.

Halstead, S.B. (2006). Dengue in the Americas and Southeast Asia: do they differ? *Revista Panamerican de Salud Publica,* 6: 407–415.

Herrera-Basto, E., Prevots, D.R., Zarate, M.L., Silva, J.L., Sepulveda-Amore, J. (1992). First reported outbreak of classical dengue fever at 1700 meters above sea level in Guerrero State, Mexico, June 1988. *American Journal of Tropical Medicine and Hygiene,* 46: 649–635.

Hopp, M.J., Foley, J.A. (2001). Global-scale relationships between climate and the dengue fever vector, *Aedes aegypti. Climate Change* 48: 441–463.

Intergovernmental Panel on Climate Change (IPCC). (2007). *Climate Change 2007: The Physical Science Basis.* Contribution of Working Group I to the Fourth Assessment Report of the Intergovernmental Panel on Climate Change. Cambridge, UK: Cambridge University Press.

Jetten, T.H., Focks, D.A. (1997). Potential changes in the distribution of dengue transmission under climate warming. *American Journal of Tropical Medicine and Hygiene,* 57: 285–297.

Johansson, M.A., Dominici, F., Glass, G.E, (2009). Local and global effects of climate on dengue transmission in Puerto Rico. *PLoS Neglected Tropical Diseases,* 3: e382. doi:10.1371/journal.pntd.0000382

Khormi, M.H., Kumar, L. (2011). Examples of using spatial information technologies for mapping and modelling mosquito-borne diseases based on environmental, climatic, socioeconomic factors and different spatial statistics, temporal risk indices and spatial analysis: a review. *Journal of Food, Agriculture and Environment,* 9: 41–49.

Khormi, H.M., Kumar, L., Elzahrany, R. (2011). Describing and analyzing the association between meteorological variables and adult *Aedes aegypti* mosquitoes. *Journal of Food, Agriculture and Environment,* 9: 954–959.

Kriticos, D.J., Webber, B.L., Leriche, A., Ota, N., Macadam, I., Bathols, J., Scott, J. (2011). CliMond: global high-resolution historical and future scenario climate surfaces for bioclimatic modelling. *Methods in Ecology and Evolution*, 53–64.

Lee, H.L, Rohani, A. (2005). Transovarial transmission of dengue virus in *Aedes aegypti* and *Aedes albopictus* in relation to dengue outbreak in an urban area in Malaysia. *Dengue Bulletin*, 29: 106–111.

Li, C.F., Lim, T.W., Han, L.L., Fang, R. (1985) Rainfall, abundance of *Aedes aegypti* and dengue infection in Selangor, Malaysia. *Southeast Asian Journal of Tropical Medicine and Public Health*, 16(4): 560–568.

Lifson, A.R. (1996). Mosquitoes, models, and dengue. *Lancet*, 347: 1201–1202.

Lounibos, LP. (2002). Invasions by insect vectors of human disease. *Annual Review of Entomology*, 47: 233–266.

Macfie, J.W.S. (1920). Heat and *Stegomyia fasciata*, short exposures to raised temperatures. *Annals of Tropical Medicine and Parasitology*, 14: 73–82.

Malavige, G.N., Fernando, S., Fernando, D.J., Seneviratne, S.L. (2004). Dengue viral infections. *Postgraduate Medical Journal*, 80: 588–601.

Martens, W.J.M., Jetten, T.H., Rotmans, J., Niessen, L.W. (1995). Climate change and vector-borne diseases. *Global Environmental Change*, 5: 195–209.

Medlock, J.M., Snow, K.R., Leach, S. (2005). Potential transmission of West Nile virus in the British Isles: an ecological review of candidate mosquito bridge vectors. *Medical and Veterinary Entomology*, 19: 2–21.

Moore, C.G. (1999). *Aedes albopictus* in the United States: current status and prospects for further spread. *Journal of the American Mosquito Control Association*, 15: 221–227.

Moore, C.G., Mitchell, C.J. (1997). *Aedes albopictus* in the United States: ten-year presence and public health implications. *Emerging Infectious Diseases*, 3: 329–334.

Ooi, E.E., Goh, K.T., Gubler, D.J. (2006) Dengue prevention and 35 years of vector control in Singapore. *Emerging Infectious Diseases*, 12: 887–893. doi:10.3201/10.3201/eid1206.051210

Patz, J.A., Martens, W.J.M., Focks, D.A., Jetten, T.H. (1998). Dengue fever epidemic potential as projected by general circulation models of global climate change. *Environmental Health Perspectives,* 106: 147–153.

Poutsma, J., Loomans, A.J.M., Aukema, B., Heijerman, T. (2008). Predicting the potential geographical distribution of the harlequin ladybird, *Harmonia axyridis*, using the CLIMEX model. *BioControl,* 53: 103–125.

Rahmstorf, S., Cazenave, A., Church, J.A., Hansen, J.E., Keeling, R.F., Parker, D.E., Somerville, R.C.J. (2007). Recent climate observations compared to projections. *Science,* 316(5825): 709.

Reiter, P. (1988). Weather, vector biology, and arboviral recrudescence. In: Monath, T.P. (Ed.), *The Arboviruses: Epidemiology and Ecology.* Vol. 1. Boca Raton, FL: CRC Press; 245–255.

Reiter, P., Gubler, D.J. (1997). Surveillance and control of urban dengue vectors. In: Gubler, D.J., Kuno, G. (Eds.), *Dengue and Dengue Hemorrhagic Fever.* Wallingford, UK: CAB International; 425–462.

Rogers, D.J., Wilson, A.J., Hay, S.I., Graham, A.J. (2006). The global distribution of yellow fever and dengue. *Advances in Parasitology,* 62: 181–220. doi:10.1016/S0065-308X(05)62006-4

Rudnick, A. (1986). Dengue fever epidemiology in Malaysia 1901–1980. In: Rudnick, A., Lim, T.W. (Eds.), *Dengue Fever Studies in Malaysia.* Bulletin No. 23. Kuala Lumpur, Malaysia: Institute of Medical Research; 9–38.

Siler, J.F., Hall, M.W., Hitchens, A.P. (1926). Dengue: its history, epidemiology, mechanisms of transmission, etiology, clinical manifestations, immunity and prevention. *Philippine Journal of Science,* 29: 1–304.

Skae, F.M.T. (1902). Dengue fever in Penang. *British Medical Journal,* 2: 1581–1582.

Smith, C.E.G. (1956). A localized outbreak of dengue fever in Kuala Lumpur: epidemiological and clinical aspects. *Medical Journal of Malaya,* 10: 289–303.

Suarez, M.F., Nelson, M.J. (1981). Registro de Altitud del *Aedes aegypti* en Colombia [Records of the altitude of *Aedes aegypti* in Colombia]. *Biomedica,* 1: 225.

Sumarmo, W.H., Jahja, E., Gubler, D.J., Suharyono, W., Sorensen, K. (1983). Clinical observations on virologically confirmed fatal dengue infections in Jakarta, Indonesia. *Bulletin of the World Health Organisation,* 61: 693–701.

Suppiah, R., Hennessy, K.J., Whetton, P.H., McInnes, K., Macadam, I., Bathols, J., Ricketts, J., Page, C.M. (2007). Australian climate change projections derived from simulations performed for the IPCC 4th assessment report. *Australian Meteorological Magazine*, 56: 131–152.

Sutherst, R.W., Bourne, A.S. (2009). Modelling non-equilibrium distributions of invasive species: a tale of two modelling paradigms. *Biological Invasions,* 11: 1231–1237.

Sutherst, R.W., Maywald, G., Kriticos, D.J. (2007). *CLIMEX Version 3: User's Guide.* South Yarra, Australia: Hearne Scientific.

Takken, W., Knols, B.G.J. (Eds.). (2007). *Emerging Pests and Vector-Borne Diseases in Europe: Ecology and Control of Vector-Borne Diseases.* Wageningen, the Netherlands: Wageningen Academic.

Taylor, S., Kumar, L., Reid, N. (2012a). Impacts of climate change and land-use on the potential distribution of an invasive weed: a case study of *Lantana camara* in Australia. *Weed Research*, 52(5): 391–401.

Taylor, S., Kumar, L., Reid, N., Kriticos, D.J. (2012b). Climate change and the potential distribution of an invasive shrub, *Lantana camara* L. *PLoS One,* 7: e35565. doi:10.1371/journal.pone.0035565

United Nations (UN). (2004). *World Population to 2300.* New York: United Nations.

Wallis, R.C. (2005). A GIS Model for Predicting Potential "High Risk" Areas of West Nile Virus by Identifying Ideal Mosquito Breeding Habitats. MSc thesis, Mississippi State University.

Wayne, A.R., Graham, C.L. (1968). The effect of temperature and relative humidity on the flight performance of female *Aedes aegypti. Journal of Insect Physiology*, 14: 1251–1257.

Wharton, T.N., Kriticos, D.J. (2004). The fundamental and realized niche of the Monterey Pine aphid, *Essigella californica* (Essig) (Hemiptera: Aphididae): implications for managing softwood plantations in Australia. *Diversity and Distributions,* 10: 253–262.

Chapter 9

Conclusion

As has been exemplified throughout this book, the prevalence of vector-borne diseases can be accurately modelled using geographic information system (GIS) technology. Models to forecast the prevalence of vector-borne diseases may be devised for localized applications with limited resources or over much broader scales. Key advantages of GIS technology are the way in which it can manage spatial data and its ability to produce risk maps that can clearly illustrate patterns and areas requiring prioritization to combat and control vector-borne diseases.

Vector-borne diseases, and more specifically the vectors, survive and reproduce within optimal climatic envelopes and consequently are affected by changes to these conditions. Changes in environmental and climatic conditions can significantly affect the rate and distance of their transmission. The factors that most influence vector-borne diseases are temperature, humidity, and precipitation; other important factors include elevation, wind speed and direction, vegetation cover, land use, and the duration of daylight (Patz et al., 2003).

Mapping vegetation density can identify the potential habitats of vectors according to their ecology. Temperature is associated with the survival, abundance, and breeding rate of vectors and affects the transmission rate of some vector-borne diseases. Rainfall, which changes from season to season, contributes to the abundance of some vectors by altering vector habitats through affecting soil moisture, land cover, and humidity. The presence of water bodies and vegetation has been shown to be closely related to the change of vector density.

The number of preventable deaths experienced on a yearly basis because of vector-borne diseases is currently in the millions. As the population increases, environmental conditions change because of urbanization and changing land use. As climatic conditions change as a consequence of a hotter Earth, there is the possibility of vectors adapting and populations developing in regions or countries that are currently unsuitable for the vector.

At present, most of the vector-borne diseases have a fairly limited distribution and require fairly specific environmental conditions for the diseases to multiply and spread (Haile, 1989). As previously mentioned, the malaria parasite is highly infectious and capable of decimating rural communities and animals. Currently, the malarial parasite is restricted to tropical and subtropical environments, but as precipitation and temperature levels across the world undergo changes, it is possible that these changes will directly affect the behaviour and geographical distribution of the vectors of this disease, which could indirectly alter the availability of suitable breeding sites around the world (Sutherst, 1998).

The use of GISs to model and simulate the prevalence of vectors and vector-borne diseases provides health professionals with the opportunity to predict future distributions and develop management programs to combat and eradicate these diseases.

The geographic aspect of a GIS and its ability to be used in conjunction with spatial analysis techniques make it an obvious tool in the management of vector-borne diseases (Rogers and Randolph, 1993). This has been demonstrated in many studies that used analysis and modelling of epidemiological data, vector data, and environmental and climate variable data to map disease distribution and abundance and high disease risk areas and predict likely future distribution and incidences.

A wide range of models and methods can be used to find areas with high risk of vector-borne diseases through the utilization of GIS capabilities. These models take into account a diverse set of variables, ranging from environmental, demographic, disease occurrence, vector distribution, and habitat variables. Spatial analysis, such as distance measures, interpolations, buffering, map overlay analysis, fuzzy analysis, cluster analysis, and many other techniques, are used in GIS platforms. Vector habitats, abundance, and larvae and adult densities are assessed in conjunction with surrounding environmental conditions to determine correlations between the vector and the environment. Land use and land cover changes over time, along with vector habitats and abundance, are mapped using GISs to identify causes of increased rates of transmission. With effective use of these technologies, although the vectors themselves cannot be identified, the environmental parameters in which they thrive can be effectively used to map, model, and subsequently provide valuable information for the effective management and control of vector-borne diseases.

An important area that should lead to greater use of GIS-based modelling of vector-borne diseases is the development of low-cost, effective GIS mapping and modelling techniques that can be utilized in developing countries that often have high incidences of vector-borne diseases yet limited resources to manage them (Booman et al., 2000; Chang et al., 2009; Martin et al., 2002). GISs are not cheap and are not readily available to many resource managers. Many personnel involved in vector-borne disease management do not have the necessary expertise in using a complex GIS. Easy-to-use systems that can integrate data from multiple sources will invariably lead to greater use of these for disease mapping and control. The development of Google Earth, for example, has enabled many millions of people to better understand the environment around them. Satellite images can now be viewed by nonexperts. Even though these systems currently do not allow for modelling or overlay of other complex data, it will not be long before ordinary users will be able to select multiple layers at a time to map and visualise extents of out- breaks and determine the direction and speed of disease spread. Future directions and advances in GISs and related technology are predicted to be in the development of methods that allow for modelling and mapping across different scales of space and time because currently GISs and related techniques are constrained by data collected to specific measures of space and time.

The primary purpose of disease surveillance is to improve the timeliness of response to disease threats to the population. Vectors and the diseases they propagate are not bound by country or administrative boundaries, yet the bulk of the data collected for the surveillance of vector-borne diseases, and most other diseases for that matter, are constrained by country or administrative boundaries. There is no universally agreed framework or standard for data collection and sharing. Districts within the same country often have different rules and standards, even for disease incidence data at clinics and hospitals. Different countries record and report the same data differently and at different spatial scales. This results in a large proportion of data being of little use in modelling exercises.

Barriers to data sharing are still a major issue in many stud- ies. If anything, easy access to much of the data is becoming harder now compared to several decades ago, even though much more and more detailed data is collected these days; this is because of privacy and confidentiality issues. In many cases, even if no confidentiality issues exist, the data is not shared because of legal issues and the culture of legal action for monetary benefit at every opportunity, especially in countries such as the United States and Australia. It is imperative that better international cooperation protocols be developed in the field of disease data sharing so that disease surveillance models can be developed to cover broader regions and to improve transmission forecasting.

The increasing popularity of air travel has substantially increased the ability and opportunity for vector-borne diseases to spread quickly across multiple countries, thereby having the potential to turn relatively localized infections into global outbreaks. These risks can be reduced or eliminated if data on outbreaks of major diseases are available easily and in a timely manner, thereby enabling at-risk countries to enact control and management protocols before the vectors and diseases reach their borders.

References

Booman, M., Durrheim, D.N., La Grange, K., Martin, C., Mabuza, A.M., Zitha, A., Mbokazi, F.M., Fraser, C., Sharp, B.L. (2000). Using a geographical information system to plan a malaria control programme in South Africa. *Bulletin of the World Health Organization*, 78: 1438–1444.

Chang, A.Y., Parrales, M.E., Jimenez, J., Sobieszczyk, M.E., Hammer, S.M., Copenhaver D.J., Kulkarni, R.P. (2009). Combining Google Earth and GIS mapping technologies in a dengue surveillance system for developing countries. *International Journal of Health Geographics*, 8: 49. doi:10.1186/1476-072X-8-49

Haile, D.G. (1989) Computer simulation of the effects of changes in weather patterns on vector-borne disease transmission. In: Smith, J.B., Tirpak, D.A. (Eds.), *The Potential Effects of Global Climate Change in the United States*. Document 230-05-89-057, Appendix G. Washington, DC: US Environmental Protection Agency.

Martin, C., Curtis, B., Fraser, C., Sharp, B. (2002). The use of a GIS-based malaria information system for malaria research and control in South Africa. *Health and Place*, 8: 227–236.

Patz, J.A., Githeko, A.K., McCarty, J.P., Hussein, S., Confalonieri, U., de Wet, N. (2003) Climate change and infectious diseases. In: McMichael, A.J., Campbell-Lendrum, D.H., Corvalan, C.F., Ebi, K.L., Githeko, A., Scheraga, J.D., Woodward, A. (Eds.), *Climate Change and Human Health: Risks and Responses*. Geneva, Switzerland: World Health Organization; 103–132.

Rogers, D.J., Randolph, S.E. (1993). Distribution of tsetse and ticks in Africa: past, present, and future. *Parasitology Today*, 9: 266–271.

Sutherst, R.W. (1998) Implications of global change and climate variability for vector-borne diseases: generic approaches to impact assessments. *International Journal for Parasitology*, 28: 935–945.

Index